GALILEO GALILEI

GREAT ASTRONOMERS

ROBERT STAWELL BALL

GIFT *Certificate*

TO:

FROM:

DATE:

Would you like to buy a copy of

GALILEO GALILEI?

PLEASE VISIT:

http://www.diamondbooks.ca

GALILEO GALILEI

GREAT ASTRONOMERS

BY

ROBERT STAWELL BALL

D.Sc, LL.D., F.R.S.

(July 1, 1840 – November 25, 1913)

LOWNDEAN PROFESSOR OF ASTRONOMY AND GEOMETRY
IN THE UNIVERSITY OF CAMBRIDGE

DIAMOND™
BOOKS
www.diamondbooks.ca

TORONTO, CANADA – 2017

DIAMOND BOOKS - CANADA

DIAMOND BOOKS

Toronto, ON, CANADA

http://www.diamondbooks.ca

BIBLIOGRAPHIC INFORMATION

'GALILEO GALILEI'
Is an extract or a chapter from the book
"GREAT ASTRONOMERS" by Robert Stawell Ball.

FIRST PUBLISHED IN: 1896.

PUBLISHED IN CANADA

Published in Canada by DIAMOND BOOKS - CANADA, an imprint of DIAMOND PUBLISHERS - http://www.diamondpublishers.com

FIRST EDITION: MARCH, 2017.

PAPERBACK EDITION : ISBN: 978-1-988357-34-8

PRINTED IN CANADA

PREFACE
TO THE ORIGINAL EDITION

It has been my object in these pages to present the life of each astronomer in such detail as to enable the reader to realise in some degree the man's character and surroundings; and I have endeavoured to indicate as clearly as circumstances would permit the main features of the discoveries by which he has become known.

There are many types of astronomers—from the stargazer who merely watches the heavens, to the abstract mathematician who merely works at his desk; it has, consequently, been necessary in the case of some lives to adopt a very different treatment from that which seemed suitable for others.

ROBERT STAWELL BALL.

The Observatory, Cambridge.
October, 1895.

NOTE: The full-length of 'Preface' can be found in its original edition of the book "GREAT ASTRONOMERS".

INTRODUCTION

Of all the natural sciences there is not one which offers such sublime objects to the attention of the inquirer as does the science of astronomy. From the earliest ages the study of the stars has exercised the same fascination as it possesses at the present day. Among the most primitive peoples, the movements of the sun, the moon, and the stars commanded attention from their supposed influence on human affairs. The practical utilities of astronomy were also obvious in primeval times. Maxims of extreme antiquity show how the avocations of the husbandman are to be guided by the movements of the heavenly bodies. The positions of the stars indicated the time to plough, and the time to sow. To the mariner who was seeking a way across the trackless ocean, the heavenly bodies .offered the only reliable marks by which his path could be guided. There was, accordingly, a stimulus both from intellectual curiosity and from practical necessity to follow the movements of the stars. Thus began a search for the causes of the ever-varying phenomena which the heavens display.

- ROBERT STAWELL BALL

Many of the earliest discoveries are indeed prehistoric. The great diurnal movement of the heavens, and the annual revolution of the sun, seem to have been known in times far more ancient than those to which any human monuments can be referred. The acuteness of the early observers enabled them to single out the more important of the wanderers which we now call planets. They saw that the star-like objects, Jupiter, Saturn, and Mars, with the more conspicuous Venus, constituted a class of bodies wholly distinct from the fixed stars among which their movements lay, and to which they bear such a superficial resemblance. But the penetration of the early astronomers went even further, for they recognized that Mercury also belongs to the same group, though this particular object is seen so rarely. It would seem that eclipses and other phenomena were observed at Babylon from a very remote period, while the most ancient records of celestial observations that we possess are to be found in the Chinese annals. The study of astronomy, in the sense in which we understand the word, may be said to have commenced under the reign of the Ptolemies at Alexandria. The most famous name in the science of this period is that of Hipparchus, who lived and worked at Rhodes about the year 160 B.C. It was his splendid investigations that first wrought the observed facts into a coherent branch of knowledge. He recognized the primary obligation which lies on the student of the heavens to compile as complete an inventory as possible of the objects which are there to be found. Hipparchus accordingly commenced by undertaking, on a small scale, a task exactly similar to that on which modern astronomers, with all available appliances of meridian circles, and photographic telescopes, are constantly engaged at the present day. He compiled a catalogue of the principal fixed stars, which is of special value to astronomers, as being the earliest work of its kind which has been handed down. He also studied the movements of the sun and the moon, and framed theories to account for the incessant changes which he saw in progress. He found a much more difficult problem in his attempt to interpret satisfactorily the complicated movements of the planets. With the view of constructing a theory which should give some coherent account of the subject, he made many observations of the places of these wandering stars. How great were the advances which Hipparchus accomplished may be appreciated if we reflect that, as a preliminary task to his more purely astronomical labours, he had to invent that branch of mathematical science by which alone the problems he proposed could be solved. It was for this purpose that he devised the indispensable method of calculation which we now know so well as trigonometry. Without the aid rendered by this beautiful art it would have been impossible for any n ally important advance in astronomical calculation to have been effected.

But the discovery which shows, beyond all others, that Hipparchus possessed one of the master-minds of all time was the detection of that remarkable celestial movement known as the precession of the equinoxes. The inquiry which conducted to this discovery involved a most profound investigation, especially when it is remembered that in the days of Hipparchus the means of observation of the heavenly bodies were only of the rudest description, and the available observations of earlier dates were extremely scanty. We can but look with astonishment on the genius of the man who, in spite of such difficulties, was able to detect such a phenomenon as the precession, and to exhibit its actual magnitude. I shall endeavour to explain the nature of this singular celestial movement, for it may be said to offer the first instance in the history of science in which we find that combination of accurate observation with skilful interpretation, of which, in the subsequent development of astronomy, we have so many splendid examples.

The word equinox implies the condition that the night is equal to the day. To a resident on the equator the night is no doubt equal to the day at all times in the year, but to one who lives on any other part of the earth, in either hemisphere, the night and the day are not generally equal. There is, however, one occasion in spring, and another in autumn, on which the day and the night are each twelve hours at all places on the earth. When the night and day are equal in spring, the point which the sun occupies on. the heavens is termed the vernal equinox. There is similarly another point in which the sun is situated at the time of the autumnal equinox. In any investigation of the celestial movements the positions of these two equinoxes on the heavens are of primary importance, and Hipparchus, with the instinct of genius, perceived their significance, and commenced to study them. It will be understood that we can always define the position of a point on the sky with reference to the surrounding stars. No doubt we do not see the stars near the sun when the sun is shining, but they are there nevertheless. The ingenuity of Hipparchus enabled him to determine the positions of each of the two equinoxes relatively to the stars which lie in its immediate vicinity. After examination of the celestial places of these points at different periods, he was led to the conclusion that each equinox was moving relatively to the stars, though that movement was so slow that twenty five thousand years would necessarily elapse before a complete circuit of the heavens was accomplished. Hipparchus traced out this phenomenon, and established it on an impregnable basis, so that all astronomers have ever since recognised the precession of the equinoxes) as one of the fundamental facts of astronomy. Not until nearly two thousand years after Hipparchus had made this splendid discovery was the explanation of its cause given by Newton.

From the days of Hipparchus down to the present hour the science of astronomy has steadily grown. One great observer after another has appeared from time to time, to reveal some new phenomenon with regard to the celestial bodies or their movements, while from time to time one commanding intellect after another has arisen to explain the true import of the facts of observations. The history of astronomy thus becomes inseparable from the history of the great men to whose labours its development is due. In the ensuing chapters we have endeavoured to sketch the lives and the work of the great philosophers, by whose labours the science of astronomy has been created.

We shall commence with Ptolemy, who, after the foundations of the science had been laid by Hipparchus, gave to astronomy the form in which it was taught throughout the Middle Ages. We shall next see the mighty revolution in our conceptions of the universe which are associated with the name of Copernicus. We then pass to those periods illumined by the genius of Galileo and Newton, and afterwards we shall trace the careers of other more recent discoverers, by whose industry and genius the boundaries of human knowledge have been so greatly extended. Our history will be brought down late enough to include some of the illustrious astronomers who laboured in the generation which has just passed away.

- ROBERT STAWELL BALL

TABLE OF CONTENTS

CHAPTERS

**An original Chapter from the book:*
"GREAT ASTRONOMERS", by Robert Stawell Ball.

CHAPTER

4

GREAT ASTRONOMERS

Galileo Galilei

(February 15, 1564 – January 8, 1642)

Galileo Galilei
(February 15, 1564 – January 8, 1642)

CHAPTER 4

GREAT ASTRONOMERS: GALILEO GALILEI

AMONG the ranks of the great astronomers it would be difficult to find one whose life presents more interesting features and remarkable vicissitudes than does that of Galileo. We may consider him as the patient investigator and brilliant discoverer. We may consider him in his private relations, especially to his daughter, Sister Maria Celeste, a woman of very remarkable character; and we have also the pathetic drama at the close of Galileo's life, when the philosopher drew down upon himself the thunders of the Inquisition.

- ROBERT STAWELL BALL

The materials for the sketch of this astonishing man are sufficiently abundant. We make special use in this place of those charming letters which his daughter wrote to him from her convent home. More than a hundred of these have been preserved, and it may well be doubted whether any more beautiful and touching series of letters addressed to a parent by a dearly loved child have ever been written. An admirable account of this correspondence is contained in a little book entitled "The Private Life of Galileo," published anonymously by Messrs. Macmillan in 1870, and I have been much indebted to the author of that volume for many of the facts contained in this chapter.

Galileo was born at Pisa, on 18th February, 1564. He was the eldest son of Vincenzo de' Bonajuti de' Galilei, a Florentine noble. Notwithstanding his illustrious birth and descent, it would seem that the home in which the great philosopher's childhood was spent was an impoverished one. It was obvious at least that the young Galileo would have to be provided with some profession by which he might earn a livelihood. From his father he derived both by inheritance and by precept a keen taste for music, and it appears that he became an excellent performer on the lute. He was also endowed with considerable artistic power, which he cultivated diligently. Indeed, it would seem that for some time the future astronomer entertained the idea of devoting himself to painting as a profession. His father, however, decided that he should study medicine. Accordingly, we find that when Galileo was seventeen years of age, and had added a knowledge of Greek and Latin to his acquaintance with the fine arts, he was duly entered at the University of Pisa.

Here the young philosopher obtained some inkling of mathematics, whereupon he became so much interested in this branch of science, that he begged to be allowed to study geometry. In compliance with his request, his father permitted a tutor to be engaged for this purpose; but he did so with reluctance, fearing that the attention of the young student might thus be withdrawn from that medical work which was regarded as his primary occupation. The event speedily proved that these anxieties were not without some justification. The propositions of Euclid proved so engrossing to Galileo that it was thought wise to avoid further distraction by terminating the mathematical tutor's engagement. But it was too late for the desired end to be attained. Galileo had now made such progress that he was able to continue his geometrical studies by himself. Presently he advanced to that famous 47th proposition which won his lively admiration, and on he went until he had mastered the six books of Euclid, which was a considerable achievement for those days.

The diligence and brilliance of the young student at Pisa did not, however, bring him much credit with the University authorities. In those days the doctrines of Aristotle were regarded as the embodiment of all human-wisdom in natural science as well as in everything else. It was regarded as the duty of every student to learn Aristotle off by heart, and any disposition to doubt or even to question the doctrines of the venerated teacher was regarded as intolerable presumption. But young Galileo had the audacity to think for himself about the laws of nature. He would not take any assertion of fact on the authority of Aristotle when he had the means of questioning nature directly as to its truth or falsehood. His teachers thus came to regard him as a somewhat misguided youth, though they could not but respect the unflagging industry with which he amassed all the knowledge he could acquire.

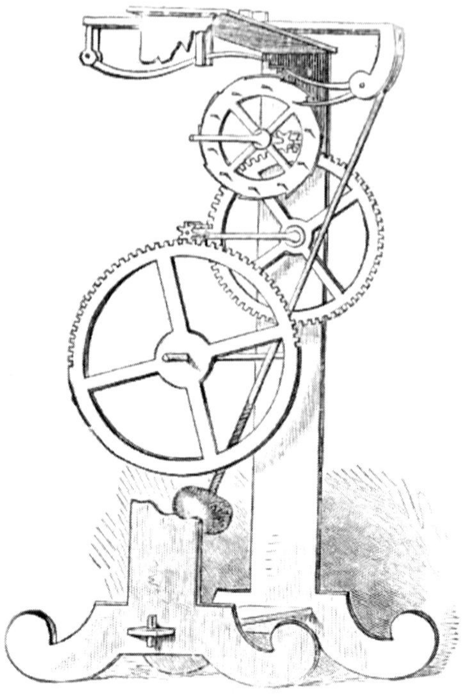

Galileo's Pendulum.

We are so accustomed to the use of pendulums in our clocks, that perhaps we do not often realise that the introduction of this method of regulating time-pieces was really a notable invention worthy the fame of the great astronomer to whom it was due. It appears that sitting one day in the Cathedral of Pisa, Galileo's attention became concentrated on the swinging of a chandelier which hung from the ceiling. It struck him as a significant point, that whether the arc through which the pendulum oscillated was a long one or a short one, the time occupied in each vibration was sensibly the same. This suggested to the thoughtful observer that a pendulum would afford the means by which a time-keeper might be controlled, and accordingly Galileo constructed for the first time a clock on this principle. The immediate object sought in this apparatus was to provide a means of aiding physicians in counting the pulses of their patients.

The talents of Galileo having at length extorted due recognition from the authorities, he was appointed, at the age of twenty-five, Professor of Mathematics at the University of Pisa. Then came the time when he felt himself strong enough to throw down the gauntlet to the adherents of the old philosophy. As a necessary part of his doctrine on the movement of bodies, Aristotle had asserted that the time occupied by a stone in falling depends upon its weight, so that the heavier the stone the less time would it require to fall from a certain height to the earth. It might have been thought that a statement so easily confuted by the simplest experiments could never have maintained its position in any accepted scheme of philosophy. But Aristotle had said it, and to anyone who ventured to express a doubt the ready sneer was forthcoming, "Do you think yourself a cleverer man than Aristotle?" Galileo determined to demonstrate in the most emphatic manner the absurdity of a doctrine which had for centuries received the sanction of the learned. The summit of the Leaning Tower of Pisa offered a highly dramatic site for the great experiment. The youthful professor let fall from the overhanging top a large heavy body and a small light body simultaneously. According to Aristotle the large body ought to have reached the ground much sooner than the small one, but such was found not to be the case. In the sight of a large concourse of people the simple fact was demonstrated that the two bodies fell side by side, and reached the ground at the same time. Thus the first great step was taken in the overthrow of that preposterous system of unquestioning adhesion to dogma, which had impeded the development of the knowledge of nature for nearly two thousand years.

This revolutionary attitude towards the ancient beliefs was not calculated to render Galileo's relations with the University authorities harmonious. He had also the misfortune to make enemies in other quarters. Don Giovanni de Medici, who was then the Governor of the Port of Leghorn, had designed some contrivance by which he proposed to pump out a dock. But Galileo showed up the absurdity of this enterprise in such an aggressive manner that Don Giovanni took mortal offence, nor was he mollified when the truths of Galileo's criticisms were abundantly verified by the total failure of his ridiculous invention. In various ways Galileo was made to feel his position at Pisa so unpleasant that he was at length compelled to abandon his chair in the University. The active exertions of his friends, of whom Galileo was so fortunate as to have had throughout his life an abundant supply, then secured his election to the Professorship of Mathematics at Padua, whither he went in 1592.

It was in this new position that Galileo entered on that marvellous career of investigation which was destined to revolutionize science. The zeal with which he discharged his professorial duties was indeed of the most unremitting character. He speedily drew such crowds to listen to his discourses on Natural Philosophy that his lecture-room was filled to overflowing. He also received many private pupils in his house for special instruction. Every moment that could be spared from these labours was devoted to his private study and to his incessant experiments.

Like many another philosopher who has greatly extended our knowledge of nature, Galileo had a remarkable aptitude for the invention of instruments designed for philosophical research. To facilitate his practical work, we find that in 1599 he had engaged a skilled workman who was to live in his house, and thus be constantly at hand to try the devices for ever springing from Galileo's fertile brain. Among the earliest of his inventions appears to have been the thermometer, which he constructed in 1602. No doubt this apparatus in its primitive form differed in some respects from the contrivance we call by the same name. Galileo at first employed water as the agent, by the expansion of which the temperature was to be measured. He afterwards saw the advantage of using spirits for the same purpose. It was not until about half a century later that mercury came to be recognised as the liquid most generally suitable for the thermometer.

The time was now approaching when Galileo was to make that mighty step in the advancement of human knowledge which followed on the application of the telescope to astronomy. As to how his idea of such an instrument originated, we had best let him tell us in his own words. The passage is given in a letter which he writes to his brother-in-law, Landucci.

"I write now because I have a piece of news for you, though whether you will be glad or sorry to hear it I cannot say; for I have now no hope of returning to my own country, though the occurrence which has destroyed that hope has had results both useful and honourable. You must know, then, that two months ago there was a report spread here that in Flanders someone had presented to Count Maurice of Nassau a glass manufactured in such a way as to make distant objects appear very near, so that a man at the distance of two miles could be clearly seen. This seemed to me so marvellous that I began to think about it. As it appeared to me to have a foundation in the Theory of Perspective, I set about contriving how to make it, and at length I found out, and have succeeded so well that the one I have made is far superior to the Dutch telescope. It was reported in Venice that I had made one, and a week since I was commanded to show it to his Serenity and to all the members of the senate, to their infinite amazement. Many gentlemen and senators, even the oldest, have ascended at various times the highest bell-towers in Venice to spy out ships at sea making sail for the mouth of the harbour, and have seen them clearly, though without my telescope they would have been invisible for more than two hours. The effect of this instrument is to show an object at a distance of say fifty miles, as if it were but five miles."

The remarkable properties of the telescope

at once commanded universal attention among intellectual men. Galileo received applications from several quarters for his new instrument, of which it would seem that he manufactured a large number to be distributed as gifts to various illustrious personages.

But it was reserved for Galileo himself to make that application of the instrument to the celestial bodies by which its peculiar powers were to inaugurate the new era in astronomy. The first discovery that was made in this direction appears to have been connected with the number of the stars. Galileo saw to his amazement that through his little tube he could count ten times as many stars in the sky as his unaided eye could detect. Here was, indeed, a surprise. We are now so familiar with the elementary facts of astronomy that it is not always easy to realise how the heavens were interpreted by the observers in those ages prior to the invention of the telescope. We can hardly, indeed, suppose that Galileo, like the majority of those who ever thought of such matters, entertained the erroneous belief that the stars were on the surface of a sphere at equal distances from the observer. No one would be likely to have retained his belief in such a doctrine when he saw how the number of visible stars could be increased tenfold by means of Galileo's telescope. It would have been almost impossible to refuse to draw the inference that the stars thus brought into view were still more remote objects which the telescope was able to reveal, just in the same way as it showed certain ships to the astonished Venetians, when at the time these ships were beyond the reach of unaided vision.

Galileo's celestial discoveries now succeeded each other rapidly. That beautiful Milky Way, which has for ages been the object of admiration to all lovers of nature, never disclosed its true nature to the eye of man till the astronomer of

Padua turned on it his magic tube. The splendid zone of silvery light was then displayed as star-dust scattered over the black back-ground of the sky. It was observed that though the individual stars were too small to be seen severally without optical aid, yet such was their incredible number that the celestial radiance produced that luminosity with which every star-gazer was so familiar.

But the greatest discovery made by the telescope in these early days, perhaps, indeed, the greatest discovery that the telescope has ever accomplished, was the detection of the system of four satellites revolving around the great planet Jupiter. This phenomenon was so wholly unexpected by Galileo that, at first, he could hardly believe his eyes. However, the reality of the existence of a system of four moons attending the great planet was soon established beyond all question. Numbers of great personages crowded to Galileo to see for themselves this beautiful miniature representing the sun with its system of revolving planets.

Of course there were, as usual, a few incredulous people who refused to believe the assertion that four more moving bodies had to be added to the planetary system. They scoffed at the notion; they said the satellites may have been in the telescope, but that they were not in the sky. One sceptical philosopher is reported to have affirmed, that even if he saw the moons of Jupiter himself he would not believe in them, as their existence was contrary to the principles of common-sense!

There can be no doubt that a special significance attached to the new discovery at this particular epoch in the history of science. It must be remembered that in those days the doctrine of Copernicus, declaring that the sun, and not the earth, was the centre of the system, that the earth revolved on its axis once a day, and that it described a mighty circle round the sun once a year, had only recently been promulgated. This new view of the scheme of nature had been encountered with the most furious opposition. It may possibly have been that Galileo himself had not felt quite confident in the soundness of the Copernican theory, prior to the discovery of the satellites of Jupiter. But when a picture was there exhibited in which a number of relatively small globes were shown to be revolving around a single large globe in the centre, it seemed impossible not to feel that the beautiful spectacle so displayed was an emblem of the relations of the planets to the sun. It was thus made manifest to Galileo that the Copernican theory of the planetary system must be the true one. The momentous import of this opinion upon the future welfare of the great philosopher will presently appear.

It would seem that Galileo regarded his residence at Padua as a state of undesirable exile from his beloved

Tuscany. He had always a yearning to go back to his own country, and at last the desired opportunity presented itself. For now that Galileo's fame had become so great, the Grand Duke of Tuscany desired to have the philosopher resident at Florence, in the belief that he would shed lustre on the Duke's dominions. Overtures were accordingly made to Galileo, and the consequence was that in 1616 we find him residing at Florence, bearing the title of Mathematician and Philosopher to the Grand Duke.

Two daughters, Polissena and Virginia, and one son, Vincenzo, had been born to Galileo in Padua. It was the custom in those days that as soon as the daughter of an Italian gentleman had grown up, her future career was somewhat summarily decided. Either a husband was to be forthwith sought out, or she was to enter the convent with the object of taking the veil as a professed nun. It was arranged that the two daughters of Galileo, while still scarcely more than children, should both enter the Franciscan convent of St. Matthew, at Arcetri. The elder daughter, Polissena, took the name of Sister Maria Celeste, while Virginia became Sister Arcangela. The latter seems to have been always delicate and subject to prolonged melancholy, and she is of but little account in the narrative of the life of Galileo. But Sister Maria Celeste, though never leaving the convent, managed to preserve a close intimacy with her beloved father. This was maintained only partly by Galileo's visits, which were very irregular and were, indeed, often suspended for long intervals. But his letters to this daughter were evidently frequent and affectionate, especially in the latter part of his life. Most unfortunately, however, all his letters have been lost. There are grounds for believing that they were deliberately destroyed when Galileo was seized by the Inquisition, lest they should have been used as evidence against him, or lest they should have compromised the convent where they were received. But Sister Maria Celeste's letters to her father have happily been preserved, and most touching these letters are. We can hardly read them without thinking how the sweet and gentle nun would have shrunk from the idea of their publication.

Her loving little notes to her "dearest lord and father," as she used affectionately to call Galileo, were almost invariably accompanied by some gift, trifling it may be, but always the best the poor nun had to bestow. The tender grace of these endearing communications was all the more precious to him from the fact that the rest of Galileo's relatives were of quite a worthless description. He always acknowledged the ties of his kindred in the most generous way, but their follies and their vices, their selfishness and their importunities, were an incessant source of annoyance to him, almost to the last day of his life.

- ROBERT STAWELL BALL

On 19th December, 1625, Sister Maria Celeste writes:—

“I send two baked pears for these days of vigil. But as the greatest treat of all, I send you a rose, which ought to please you extremely, seeing what a rarity it is at this season; and with the rose you must accept its thorns, which represent the bitter passion of our Lord, whilst the green leaves represent the hope we may entertain that through the same sacred passion we, having passed through the darkness of the short winter of our mortal life, may attain to the brightness and felicity of an eternal spring in heaven.”

When the wife and children of Galileo's shiftless brother came to take up their abode in the philosopher's home, Sister Maria Celeste feels glad to think that her father has now someone who, however imperfectly, may fulfil the duty of looking after him. A graceful note on Christmas Eve accompanies her little gifts. She hopes that—

“In these holy days the peace of God may rest on him and all the house. The largest collar and sleeves I mean for Albertino, the other two for the two younger boys, the little dog for baby, and the cakes for everybody, except the spice-cakes, which are for you. Accept the goodwill which would readily do much more.”

The extraordinary forbearance with which Galileo continually placed his time, his purse, and his influence at the service of those who had repeatedly proved themselves utterly unworthy of his countenance, is thus commented on by the good nun:—

“Now it seems to me, dearest lord and father, that your lordship is walking in the right path, since you take hold of every occasion that presents itself to shower continual benefits on those who only repay you with ingratitude. This is an action which is all the more virtuous and perfect as it is the more difficult."

When the plague was raging in the neighbourhood, the loving daughter's solicitude is thus shown:—

"I send you two pots of electuary as a preventive against the plague. The one without the label consists of dried figs, walnuts, rue, and salt, mixed together with honey. A piece of the size of a walnut to be taken in the morning, fasting, with a little Greek wine."

The plague increasing still more, Sister Maria Celeste obtained, with much difficulty, a small quantity of a renowned liqueur, made by Abbess Ursula, an exceptionally saintly nun. This she sends to her father with the words :—

"I pray your lordship to have faith in this remedy. For if you have so much faith in my poor miserable prayers, much more may you have in those of such a holy person; indeed, through her merits you may feel sure of escaping all danger from the plague."

Whether Galileo took the remedy we do not know, but at all events he escaped the plague.

The Villa Arcetri.
Galileo's residence, where Milton visited him.

From Galileo's new home in Florence the telescope was again directed to the skies, and again did astounding discoveries reward the astronomer's labours. The great success which he had met with in studying Jupiter naturally led Galileo to look at Saturn. Here he saw a spectacle which was sufficiently amazing, though he failed to interpret it accurately. It was quite manifest that Saturn did not exhibit a simple circular disc like Jupiter, or like Mars. It seemed to Galileo as if the planet consisted of three bodies, a large globe in the centre, and a smaller one on each side. The enigmatical nature of the discovery led Galileo to announce it in an enigmatical manner. He published a string of letters which, when duly transposed, made up a sentence which affirmed that the planet Saturn was threefold. Of course we now know that this remarkable appearance of the planet was due to the two projecting portions of the ring. With the feeble power of Galileo's telescope, these seemed merely like small globes or appendages to the large central body.

The last of Galileo's great astronomical discoveries related to the libration of the moon. I think that the detection of this phenomenon shows his acuteness of observation more remarkably than does any one of his other achievements with the telescope. It is well known that the moon constantly keeps the same face turned towards the earth. When, however, careful measurements have been made with regard to the spots and marks on the lunar surface, it is found that there is a slight periodic variation which permits us to see now a little to the east or to the west, now a little to the north or to the south of the average lunar disc.

But the circumstances which make the career of Galileo so especially interesting from the biographer's point of view, are hardly so much the triumphs that he won as the sufferings that he endured. The sufferings and the triumphs were, however, closely connected, and it is fitting that we should give due consideration to what was perhaps the greatest drama in the history of science.

On the appearance of the immortal work of Copernicus, in which it was taught that the earth rotated on its axis, and that the earth, like the other planets, revolved round the sun, orthodoxy stood aghast. The Holy Roman Church submitted this treatise, which bore the name "De Revolutionibus Orbium Coelestium," to the Congregation of the Index. After due examina-tion it was condemned as heretical in 1615. Galileo was suspected, on no doubt excellent grounds, of entertaining the objectionable views of Copernicus. He was accordingly privately summoned before Cardinal Bellarmine on 26th February, 1616, and duly admonished that he was on no account to teach or to defend the obnoxious doctrines. Galileo was much distressed by this intimation. He felt it a serious matter to be deprived of the privilege of discoursing with his friends about the Copernican system, and of instructing his disciples in the principles of the great theory of whose truth he was perfectly convinced. It pained him, however, still more to think, devout Catholic as he was, that such suspicions of his fervent allegiance to his Church should ever have existed, as were implied by the words and monitions of Cardinal Bellarmine.

In 1616, Galileo had an interview with Pope Paul V., who received the great astronomer very graciously, and walked up and down with him in conversation for three-quarters of an hour. Galileo complained to his Holiness of the attempts made by his enemies to embarrass him with the authorities of the Church, but the Pope bade him be comforted. His Holiness had himself no doubts of Galileo's orthodoxy, and he assured him that the Congregation of the Index should give Galileo no further trouble so long as Paul V. was in the chair of St. Peter.

On the death of Paul V. in 1623, Maffeo Barberini was elected Pope, as Urban VIII. This new Pope, while a cardinal, had been an intimate friend of Galileo's, and had indeed written Latin verses in praise of the great astronomer and his discoveries. It was therefore not unnatural for Galileo to think that the time had arrived when, with the use of due circumspection, he might continue his studies and his writings, without fear of incurring the displeasure of the Church. Indeed, in 1624, one of Galileo's friends writing from Rome, urges Galileo to visit the city again, and added that—

“Under the auspices of this most excellent, learned, and benignant Pontiff, science must flourish. Your arrival will be welcome to his Holiness. He asked me if you were coming, and when, and in short, he seems to love and esteem you more than ever.”

The visit was duly paid, and when Galileo returned to Florence, the Pope wrote a letter from which the following is an extract, commending the philosopher to the good offices of the young Ferdinand, who had shortly before succeeded his father in the Grand Duchy of Tuscany.

"We find in Galileo not only literary distinction, but also the love of piety, and he is also strong in those qualities by which the pontifical good-will is easily obtained. And now, when he has been brought to this city to congratulate us on our elevation, we have very lovingly embraced him; nor can we suffer him to return to the country whither your liberality calls him, without an ample provision of pontifical love. And that you may know how dear he is to us, we have willed to give him this honour able testimonial of virtue and piety. And we further signify that every benefit which you shall confer upon him, imitating or even surpassing your father's liberality, will conduce to our gratification."

The favourable reception which had been accorded to him by Pope Urban VIII. seems to have led Galileo to expect that there might be some corresponding change in the attitude of the Papal authorities on the great question of the stability of the earth. He accordingly proceeded with the preparation of the chief work of his life, "The Dialogue of the Two Systems." It was submitted for inspection by the constituted authorities. The Pope himself thought that, if a few conditions which he laid down were duly complied with, there could be no objection to the publication of the work. In the first place, the title of the book was to be so carefully worded as to show plainly that the Copernican doctrine was merely to be regarded as an hypothesis, and not as a scientific fact. Galileo was also instructed to conclude the book with special arguments which had been supplied by the Pope himself, and which appeared to his Holiness to be quite conclusive against the new doctrine of Copernicus.

Formal leave for the publication of the Dialogue was then given to Galileo by the Inquisitor General, and it was accordingly sent to the press. It might be thought that the anxieties of the astronomer about his book would then have terminated. As a matter of fact, they had not yet seriously begun. Riccardi, the Master of the Sacred Palace, having suddenly had some further misgivings, sent to Galileo for the manuscript while the work was at the printer's, in order that the doctrine it implied might be once again examined. Apparently, Riccardi had come to the conclusion that he had not given the matter sufficient attention, when the authority to go to press had been first and, perhaps, hastily given. Considerable delay in the issue of the book was the result of these further deliberations. At last, however, in June, 1632, Galileo's great work, "The Dialogue of the Two Systems," was produced for the instruction of the world, though the occasion was fraught with ruin to the immortal author.

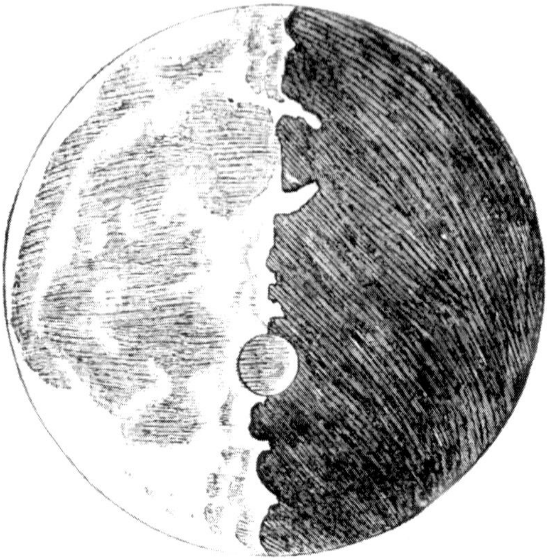

Facsimile Sketch of Lunar Surface by Galileo.

The book, on its publication, was received and read with the greatest avidity. But presently the Master of the Sacred Palace found reason to regret that he had given his consent to its appearance. He accordingly issued a peremptory order to sequestrate every copy in Italy. This sudden change in the Papal attitude towards Galileo formed the subject of a strong remonstrance addressed to the Roman authorities by the Grand Duke of Tuscany. The Pope himself seemed to have become impressed all at once with the belief that the work contained matter of an heretical description. The general interpretation put upon the book seems to have shown the authorities that they had mistaken its true tendency, notwithstanding the fact that it had been examined again and again by theologians deputed for the duty. To the communication from the Grand Duke the Pope returned answer, that he had decided to submit the book to a congregation of "learned, grave, and saintly men, "who would weigh every word in it. The views of his Holiness personally on the subject were expressed in his belief that the Dialogue contained the most perverse matter that could come into a reader's hands.

- ROBERT STAWELL BALL

The Master of the Sacred Palace was greatly blamed by the authorities for having given his sanction to its issue. He pleaded that the book had not been printed in the precise terms of the original manuscript which had been submitted to him. It was also alleged that Galileo had not adhered to his promise of inserting properly the arguments which the Pope himself had given in support of the old and orthodox view. One of these had, no doubt, been introduced, but, so far from mending Galileo's case, it had made matters really look worse for the poor philosopher. The Pope's argument had been put into the mouth of one of the characters in the Dialogue named "Simplicio." Galileo's enemies maintained that by adopting such a method for the expression of his Holiness's opinion, Galileo had intended to hold the Pope himself up to ridicule. Galileo's friends maintained that nothing could have been farther from his intention. It seems, however, highly probable that the suspicions thus aroused had something to say to the sudden change of front on the part of the Papal authorities.

On 1st October, 1632, Galileo received an order to appear before the Inquisition at Rome on the grave charge of heresy. Galileo, of course, expressed his submission, but pleaded for a respite from compliance with the summons, on the ground of his advanced age and his failing health. The Pope was, however, inexorable; he said that he had warned Galileo of his danger while he was still his friend. The command could not be disobeyed. Galileo might perform the journey as slowly as he pleased, but it was imperatively necessary for him to set forth, and at once.

On 20th January, 1633, Galileo started on his weary journey to Rome, in compliance with this peremptory summons. On 13th February he was received as the guest of Niccolini, the Tuscan ambassador, who had acted as his wise and ever-kind friend throughout the whole affair. It seemed plain that the Holy Office were inclined to treat Galileo with as much clemency and consideration as was consistent with the determination that the case against him should be proceeded with to the end. The Pope intimated that in consequence of his respect for the Grand Duke of Tuscany he should permit Galileo to enjoy the privilege, quite unprecedented for a prisoner charged with heresy, of remaining as an inmate in the ambassador's house. He ought, strictly, to have been placed in the dungeons of the Inquisition. When the examination of the accused had actually commenced, Galileo was confined, not, indeed, in the dungeons, but in comfortable rooms at the Holy Office.

By the judicious and conciliatory language of submission which Niccolini had urged Galileo to use before the Inquisitors, they were so far satisfied that they interceded with the Pope for his release. During the remainder of the trial Galileo was accordingly permitted to go back to the ambassador's, where he was most heartily welcomed. Sister Maria Celeste, evidently thinking this meant that the whole case was at an end, thus expresses herself:—

“The joy that your last dear letter brought me, and the having to read it over and over to the nuns, who made quite a jubilee on hearing its contents, put me into such an excited state that at last I got a severe attack of headache.”

In his defence Galileo urged that he had already been acquitted in 1616 by Cardinal Bellarmine, when a charge of heresy was brought against him, and he contended that anything he might now have done, was no more than he had done on the preceding occasion, when the orthodoxy of his doctrines received solemn confirmation. The Inquisition seemed certainly inclined to clemency, but the Pope was not satisfied. Galileo was accordingly summoned again on the 21st

June. He was to be threatened with torture if he did not forthwith give satisfactory explanations as to the reasons which led him to write the Dialogue. In this proceeding the Pope assured the Tuscan ambassador that he was treating Galileo with the utmost consideration possible in consequence of his esteem and regard for the Grand Duke, whose servant Galileo was. It was, however, necessary that some exemplary punishment be meted out to the astronomer, inasmuch as by the publication of the Dialogue he had distinctly disobeyed the injunction of silence laid upon him by the decree of 1616. Nor was it admissible for Galileo to plead that his book had been sanctioned by the Master of the Sacred College, to whose inspection it had been again and again submitted. It was held, that if the Master of the Sacred College had been unaware of the solemn warning the philosopher had already received sixteen years previously, it was the duty of Galileo to have drawn his attention to that fact.

On the 22nd June, 1633, Galileo was led to the great hall of the Inquisition, and compelled to kneel before the cardinals there assembled and hear his sentence. In a long document, most elaborately drawn up, it is definitely charged against Galileo that, in publishing the Dialogue, he committed the essentially grave error of treating the doctrine of the earth's motion as open to discussion. Galileo knew, so the document affirmed, that the Church had emphatically pronounced this notion to be contrary to Holy Writ, and that for him to consider a doctrine so stigmatized as having any shadow of probability in its favour was an act of disrespect to the authority of the Church which could not be overlooked. It was also charged against Galileo that in his Dialogue he has put the strongest arguments into the mouth, not of those who supported the orthodox doctrine, but of those who held the theory as to the earth's motion which the Church had so deliberately condemned.

- ROBERT STAWELL BALL

After due consideration of the defence made by the prisoner, it was thereupon decreed that he had rendered himself vehemently suspected of heresy by the Holy Office, and in consequence had incurred all the censures and penalties of the sacred canons, and other decrees promulgated against such persons. The graver portion of these punishments would be remitted, if Galileo would solemnly repudiate the heresies referred to by an abjuration to be pronounced by him in the terms laid down.

At the same time it was necessary to mark, in some emphatic manner, the serious offence which had been committed, so that it might serve both as a punishment to Galileo and as a warning to others. It was accordingly decreed that he should be condemned to imprisonment in the Holy Office during the pleasure of the Papal authorities, and that he should recite once a week for three years the seven Penitential Psalms.

Then followed that ever-memorable scene in the great hall of the Inquisition, in which the aged and infirm Galileo, the inventor of the telescope and the famous astronomer, knelt down to abjure before the most eminent and reverend Lords Cardinal, Inquisitors General throughout the Christian Republic against heretical depravity. With his hands on the Gospels, Galileo was made to curse and detest the false opinion that the sun was the centre of the universe and immovable, and that the earth was not the centre of the same, and that it moved. He swore that for the future he will never say nor write such things as may bring him under suspicion, and that if he does so he submits to all the pains and penalties of the sacred canons. This abjuration was subsequently read in Florence before Galileo's disciples, who had been specially summoned to attend.

It has been noted that neither on the first occasion, in 1616, nor on the second in 1633, did the reigning Pope sign the decrees concerning Galileo. The contention has accordingly been made that Paul V. and Urban VIII. are both alike vindicated from any technical responsibility for the attitude of the Romish Church towards the Copernican doctrines. The significance of this circumstance has been commented on in connection with the doctrine of the infallibility of the Pope.

We can judge of the anxiety felt by Sister Maria Celeste about her beloved father during these terrible trials. The wife of the ambassador Niccolini, Galileo's steadfast friend, most kindly wrote to give the nun whatever quieting assurances the case would permit. There is a renewed flow of these touching epistles from the daughter to her father. Thus she sends word—

“The news of your fresh trouble has pierced my soul with grief all the more that it came quite unexpectedly.”

And again, on hearing that he had been permitted to leave Rome, she writes:—

“I wish I could describe the rejoicing of all the mothers and sisters on hearing of your happy arrival at Siena. It was indeed most extraordinary. On hearing the news the Mother Abbess and many of the nuns ran to me, embracing me and weeping for joy and tenderness.”

The sentence of imprisonment was at first interpreted leniently by the Pope. Galileo was allowed to reside in qualified durance in the archbishop's house at Siena. Evidently the greatest pain that he endured arose from the forced separation from that daughter, whom he had at last learned to love with an affection almost comparable with that she bore to him. She had often told him that she never had any pleasure equal to that with which she rendered any service to her father. To her joy, she discovers that she can relieve him from the task of reciting the seven Penitential Psalms which had been imposed as a penance:—

- ROBERT STAWELL BALL

“I began to do this a while ago, “she writes,” and it gives me much pleasure. First, because I am persuaded that prayer in obedience to Holy Church must be efficacious; secondly, in order to save you the trouble of remembering it. If I had been able to do more, most willingly would I have entered a straiter prison than the one I live in now, if by so doing I could have set you at liberty.”

Sister Maria Celeste was gradually failing in health, but the great privilege was accorded to her of being able once again to embrace her beloved lord and master. Galileo had, in fact, been permitted to return to his old home; but on the very day when he heard of his daughter's death came the final decree directing him to remain in his own house in perpetual solitude.

Crest of Galileo's Family.

Amid the advancing infirmities of age, the isolation from friends, and the loss of his daughter, Galileo once again sought consolation in hard work. He commenced his famous dialogue on Motion. Gradually, however, his sight began to fail, and blindness was at last added to his other troubles. On January 2nd, 1638, he writes to Diodati:—

"Alas, your dear friend and servant, Galileo, has been for the last month perfectly blind, so that this heaven, this earth, this universe, which I by my marvellous discoveries and clear demonstrations have enlarged a hundred thousand times beyond the belief of the wise men of bygone ages, henceforward is for me shrunk into such a small space as is filled by my own bodily sensations."

But the end was approaching—the great philosopher, was attacked by low fever, from which he died on the 8th January, 1643.

COMPLIMENTARY MATERIAL

A BRIEF BIOGRAPHICAL SKETCH

GALILEO GALILEI

- ROBERT STAWELL BALL

Galileo Galilei
(February 15, 1564 – January 8, 1642)

COMPLIMENTARY MATERIAL

A BRIEF BIOGRAPHICAL SKETCH

GALILEO GALILEI

Galileo Galilei (15 February 1564 – 8 January 1642) was an Italian polymath: astronomer, physicist, engineer, philosopher, and mathematician.

He played a major role in the scientific revolution of the seventeenth century. His contributions to observational astronomy include the telescopic confirmation of the phases of Venus, the discovery of the four largest satellites of Jupiter (named the Galilean moons in his honour), and the observation and analysis of sunspots. Galileo also worked in applied science and technology, inventing an improved military compass and other instruments.

Galileo's championing of heliocentrism and Copernicanism was controversial during his lifetime, when most subscribed to either geocentrism or the Tychonic system. He met with opposition from astronomers, who doubted heliocentrism because of the absence of an observed stellar parallax. The matter was investigated by the Roman Inquisition in 1615, which concluded that heliocentrism was "foolish and absurd in philosophy, and formally heretical since it explicitly contradicts in many places the sense of Holy Scripture." Galileo later defended his views in Dialogue Concerning the Two Chief World Systems, which appeared to attack Pope Urban VIII and thus alienated him and the Jesuits, who had both supported Galileo up until this point. He was tried by the Inquisition, found "vehemently suspect of heresy", and forced to recant. He spent the rest of his life under house arrest. While under house arrest, he wrote one of his best-known works, Two New Sciences, in which he summarized work he had done some forty years earlier on the two sciences now called kinematics and strength of materials.

He has been called the "father of observational astronomy", the "father of modern physics", the "father of scientific method", and the "father of science".

Early Life

Galileo was born in Pisa (then part of the Duchy of Florence), Italy, on 15 February 1564, the first of six children of Vincenzo Galilei, a famous lutenist, composer, and music theorist, and Giulia (née Ammannati), who had married in 1562. Galileo became an accomplished lutenist himself and would have learned early from his father a scepticism for established authority, the value of well-measured or quantified experimentation, an appreciation for a periodic or musical measure of time or rhythm, as well as the results expected from a combination of mathematics and experiment.

Three of Galileo's five siblings survived infancy. The youngest, Michelangelo (or Michelagnolo), also became a noted lutenist and composer although he contributed to financial burdens during

Galileo's young adulthood. Michelangelo was unable to contribute his fair share of their father's promised dowries to their brothers-in-law, who would later attempt to seek legal remedies for payments due. Michelangelo would also occasionally have to borrow funds from Galileo to support his musical endeavours and excursions. These financial burdens may have contributed to Galileo's early fire to develop inventions that would bring him additional income.

When Galileo Galilei was eight, his family moved to Florence, but he was left with Jacopo Borghini for two years. He then was educated in the Camaldolese Monastery at Vallombrosa, 35 km southeast of Florence.

Name

The surname Galilei derives from the given name of an ancestor, Galileo Bonaiuti, a physician, university teacher and politician who lived in Florence from 1370 to 1450; his descendents had changed their family name from Bonaiuti (or Buonaiuti) to Galilei in his honor in the late 14th century. Galileo Bonaiuti was buried in the same church, the Basilica of Santa Croce in Florence, where about 200 years later his more famous descendant Galileo Galilei was also buried.

It was common for mid-sixteenth century Tuscan families to name the eldest son after the parents' surname - hence, Galileo Galilei was not necessarily named after his ancestor Galileo Bonaiuti. The Italian male given name "Galileo" (and thence the surname "Galilei") derives from the Latin "Galilaeus", meaning "of Galilee", a biblically significant region in Northern Israel.

The biblical roots of Galileo's name and surname were to become the subject of a famous pun. In 1614, during the Galileo affair, one of Galileo's opponents, the

Dominican priest Tommaso Caccini, delivered against Galileo a controversial and influential sermon. In it he made a point of quoting Acts 1:11, "Ye men of Galilee, why stand ye gazing up into heaven?".

Children

Despite being a genuinely pious Roman Catholic, Galileo fathered three children out of wedlock with Marina Gamba. They had two daughters, Virginia (born in 1600) and Livia (born in 1601), and a son, Vincenzo (born in 1606).

Because of their illegitimate birth, their father considered the girls unmarriageable, if not posing problems of prohibitively expensive support or dowries, which would have been similar to Galileo's previous extensive financial problems with two of his sisters. Their only worthy alternative was the religious life. Both girls were accepted by the convent of San Matteo in Arcetri and remained there for the rest of their lives. Virginia took the name Maria Celeste upon entering the convent. She died on 2 April 1634, and is buried with Galileo at the Basilica of Santa Croce, Florence. Livia took the name Sister Arcangela and was ill for most of her life. Vincenzo was later legitimised as the legal heir of Galileo and married Sestilia Bocchineri.

Career as a Scientist

Although Galileo seriously considered the priesthood as a young man, at his father's urging he instead enrolled at the University of Pisa for a medical degree. In 1581, when he was studying medicine, he noticed a swinging chandelier, which air currents shifted about to swing in larger and smaller arcs. To him it seemed, by comparison with his heartbeat, that the chandelier took the same amount of time to swing back and forth, no matter how far it was swinging. When he returned home, he set up two pendulums of equal length and swung one with a large sweep and the other with a small sweep and found that they kept time together. It was not until the work of Christiaan Huygens, almost one hundred years later, that the tautochrone nature of a swinging pendulum was used to create an accurate timepiece. Up to this point, Galileo had deliberately been kept away from mathematics, since a physician earned a higher income than a mathematician. However, after accidentally attending a lecture on geometry, he talked his reluctant father into letting him study mathematics and natural philosophy instead of medicine. He created a thermoscope, a forerunner of the thermometer, and in 1586 published a small book on the design of a hydrostatic balance he had invented (which first brought him to the attention of the scholarly world). Galileo also studied disegno, a term encompassing fine art, and in 1588 obtained the position of instructor in the Accademia delle Arti del Disegno in Florence, teaching perspective and chiaroscuro. Being inspired by the artistic tradition of the city and the works of the Renaissance artists, Galileo acquired an aesthetic mentality. While a young teacher at the Accademia, he began a lifelong friendship with the Florentine painter Cigoli, who included Galileo's lunar observations in one of his paintings.

In 1589, he was appointed to the chair of mathematics in Pisa. In 1591, his father died, and he was entrusted with the care of his younger brother Michelagnolo. In 1592, he moved to the University of Padua where he taught geometry, mechanics, and astronomy until 1610. During this period, Galileo made significant discoveries in both pure fundamental science (for example, kinematics of motion and astronomy) as well as practical applied science (for example, strength of materials and pioneering the telescope). His multiple interests included the study of astrology, which at the time was a discipline tied to the studies of mathematics and astronomy.

Galileo's beloved elder daughter, Virginia (Sister Maria Celeste), was particularly devoted to her father. She is buried with him in his tomb in the Basilica of Santa Croce, Florence.

Galileo, Kepler and theories of tides

Cardinal Bellarmine had written in 1615 that the Copernican system could not be defended without "a true physical demonstration that the sun does not circle the earth but the earth circles the sun". Galileo considered his theory of the tides to provide the required physical proof of the motion of the earth. This theory was so important to him that he originally intended to entitle his Dialogue on the Two Chief World Systems the Dialogue on the Ebb and Flow of the Sea. The reference to tides was removed from the title by order of the Inquisition.

For Galileo, the tides were caused by the sloshing back and forth of water in the seas as a point on the Earth's surface sped up and slowed down because of the Earth's rotation on its axis and revolution around the Sun. He circulated his first account of the tides in 1616, addressed to Cardinal Orsini. His theory gave the first insight into the importance of the shapes of ocean basins in the size and timing of tides; he correctly accounted, for instance, for the negligible tides halfway along the Adriatic Sea compared to those at the ends. As a general account of the cause of tides, however, his theory was a failure.

If this theory were correct, there would be only one high tide per day. Galileo and his contemporaries were aware of this inadequacy because there are two daily high tides at Venice instead of one, about twelve hours apart. Galileo dismissed this anomaly as the result of several secondary causes including the shape of the sea, its depth, and other factors. Against the assertion that Galileo was deceptive in making these arguments, Albert Einstein expressed the opinion that Galileo developed his "fascinating arguments" and accepted them uncritically out of a desire for physical proof of the motion of the Earth. Galileo dismissed the idea, held by his contemporary Johannes Kepler, that the moon caused the tides.] (Galileo also took no interest in Kepler's elliptical orbits of the planets.)

Galileo Galilei. Portrait by Leoni

Controversy over comets and The Assayer

In 1619, Galileo became embroiled in a controversy with Father Orazio Grassi, professor of mathematics at the Jesuit Collegio Romano. It began as a dispute over the nature of comets, but by the time Galileo had published The Assayer (Il Saggiatore) in 1623, his last salvo in the dispute, it had become a much wider controversy over the very nature of science itself. The title page of the book describes Galileo as philosopher and "Matematico Primario" of the Grand Duke of Tuscany.

Because The Assayer contains such a wealth of Galileo's ideas on how science should be practised, it has been referred to as his scientific manifesto. Early in 1619, Father Grassi had anonymously published a pamphlet, *An Astronomical Disputation on the Three Comets of the Year 1618*, which discussed the nature of a comet that had appeared late in November of the previous year. Grassi concluded that the comet was a fiery body which had moved along a segment of a great circle at a constant distance from the earth, and since it moved in the sky more slowly than the moon, it must be farther away than the moon.

Grassi's arguments and conclusions were criticised in a subsequent article, Discourse on the Comets, published under the name of one of Galileo's disciples, a Florentine lawyer named Mario Guiducci, although it had been largely written by Galileo himself. Galileo and Guiducci offered no definitive theory of their own on the nature of comets although they did present some tentative conjectures that are now known to be mistaken. In its opening passage, Galileo and Guiducci's Discourse gratuitously insulted the Jesuit Christopher Scheiner, and various uncomplimentary remarks about the professors of the Collegio Romano were scattered throughout the work. The Jesuits were offended, and Grassi soon replied with a polemical tract of his own, The Astronomical and Philosophical Balance, under the pseudonym Lothario

Sarsio Sigensano, purporting to be one of his own pupils.

The Assayer was Galileo's devastating reply to the *Astronomical Balance*. It has been widely recognized as a masterpiece of polemical literature, in which "Sarsi's" arguments are subjected to withering scorn. It was greeted with wide acclaim, and particularly pleased the new pope, Urban VIII, to whom it had been dedicated. In Rome, in the previous decade, Barberini, the future Urban VIII, had come down on the side of Galileo and the Lincean Academy.

Galileo's dispute with Grassi permanently alienated many of the Jesuits who had previously been sympathetic to his ideas, and Galileo and his friends were convinced that these Jesuits were responsible for bringing about his later condemnation. The evidence for this is at best equivocal, however.

Controversy over Heliocentrism

In the Christian world prior to Galileo's conflict with the Church, the majority of educated people subscribed either to the Aristotelian geocentric view that the earth was the center of the universe and that all heavenly bodies revolved around the Earth, or the Tychonic system that blended geocentrism with heliocentrism. Nevertheless, following the death of Copernicus and before Galileo, heliocentrism was relatively uncontroversial; Copernicus's work was used by Pope Gregory XIII to reform the calendar in 1582.

Opposition to heliocentrism and Galileo's writings combined religious and scientific objections and were fueled by political events. Scientific opposition came from Tycho Brahe and others, and arose from the fact that, if heliocentrism were true, an annual stellar parallax should be observed, though none was. Copernicus had correctly postulated that parallax was negligible because the stars were so distant. However, Brahe had countered that, since stars appeared to have measurable size, if the stars were that distant, they would be gigantic, and in fact far larger than the Sun or any other celestial body. In Brahe's system, by contrast, the stars were a little more distant than Saturn, and the Sun and stars were comparable in size.

Cristiano Banti's 1857 painting Galileo facing the *Roman Inquisition*

Religious opposition to heliocentrism arose from Biblical references such as Psalm 93:1, 96:10, and 1 Chronicles 16:30 which include text stating that "the world is firmly established, it cannot be moved." In the same manner, Psalm 104:5 says, "the Lord set the earth on its foundations; it can never be moved." Further, Ecclesiastes 1:5 states that "And the sun rises and sets and returns to its place."

Galileo defended heliocentrism based on his astronomical observations of 1609 (Sidereus Nuncius 1610). In December 1613 the Grand Duchess Christina of Florence confronted one of Galileo's friends and followers, Benedetto Castelli, with biblical objections to the motion of the earth. According to Maurice Finocchiaro this was done in a friendly and gracious manner, out of curiosity. Prompted by this incident, Galileo wrote a letter to Castelli in which he argued that heliocentrism was actually not contrary to biblical texts, and that the bible was an authority on faith and morals, not on science. This letter was not published, but circulated widely.

By 1615 Galileo's writings on heliocentrism had been submitted to the Roman Inquisition by Father Niccolo Lorini, who claimed that Galileo and his followers were attempting to reinterpret the Bible, which was seen as a violation of the Council of Trent and looked dangerously like Protestantism. Lorini specifically cited Galileo's letter to Castelli. Galileo went to Rome to defend himself and his Copernican and biblical ideas. At the start of 1616, Monsignor Francesco Ingoli initiated a debate with Galileo, sending him an essay disputing the Copernican system. Galileo later stated that he believed this essay to have been instrumental in the action against Copernicanism that followed. According to Maurice Finocchiaro, Ingoli had probably been commissioned by the Inquisition to write an expert opinion on the controversy, and the essay provided the "chief direct basis" for the Inquisition's actions. The essay focused on eighteen physical and mathematical arguments against heliocentrism. It borrowed primarily from the arguments of Tycho Brahe, and it notedly mentioned Brahe's argument that heliocentrism required the stars to be much larger than the Sun. Ingoli wrote that the great distance to the stars in the heliocentric theory "clearly proves ... the fixed stars to be of such size, as they may surpass or equal the size of the orbit circle of the Earth itself." The essay also included four theological arguments, but Ingoli suggested Galileo focus on the physical and mathematical arguments, and he did not mention Galileo's biblical ideas. In February 1616, an Inquisitorial commission declared heliocentrism to be "foolish and absurd in philosophy, and formally heretical since it explicitly contradicts in many places the sense of Holy Scripture." The Inquisition found that the idea of the Earth's movement "receives the same judgement in philosophy and... in regard to theological truth it is at least erroneous in faith".

Pope Paul V instructed Cardinal Bellarmine to deliver this finding to Galileo, and to order him to abandon the opinion that heliocentrism was physically true. On 26 February, Galileo was called to Bellarmine's residence and ordered:

... to abandon completely... the opinion that the sun stands still at the center of the world and the earth moves, and henceforth not to hold, teach, or defend it in any way whatever, either orally or in writing.

The decree of the Congregation of the Index banned Copernicus's De Revolutionibus and other heliocentric works until correction. Bellarmine's instructions did not prohibit Galileo from discussing heliocentrism as a mathematical and philosophic idea, so long as he did not advocate for its physical truth.

For the next decade, Galileo stayed well away from the controversy. He revived his project of writing a book on the subject, encouraged by the election of Cardinal Maffeo Barberini as Pope Urban VIII in 1623. Barberini was a friend and admirer of Galileo, and had opposed the condemnation of Galileo in 1616. Galileo's resulting book, *Dialogue Concerning the Two Chief World Systems*, was published in 1632, with formal authorization from the Inquisition and papal permission.

Earlier, Pope Urban VIII had personally asked Galileo to give arguments for and against heliocentrism in the book, and to be careful not to advocate heliocentrism. He made another request, that his own views on the matter be included in Galileo's book. Only the latter of those requests was fulfilled by Galileo.

Whether unknowingly or deliberately, Simplicio, the defender of the Aristotelian geocentric view in Dialogue Concerning the Two Chief World Systems, was often caught in his own errors and sometimes came across as a fool. Indeed, although Galileo states in the preface of his book that the character is named after a famous Aristotelian philosopher (Simplicius in Latin, "Simplicio" in Italian), the name "Simplicio" in Italian also has the connotation of "simpleton". This portrayal of Simplicio made Dialogue Concerning the Two Chief World Systems appear as an advocacy book: an attack on Aristotelian geocentrism and defence of the Copernican theory. Unfortunately for his relationship with the Pope, Galileo put the words of Urban VIII into the mouth of Simplicio.

Most historians agree Galileo did not act out of malice and felt blindsided by the reaction to his book. However, the Pope did not take the suspected public ridicule lightly, nor the Copernican advocacy.

Galileo had alienated one of his biggest and most powerful supporters, the Pope, and was called to Rome to defend his writings in September 1632. He finally arrived in February 1633 and was brought before inquisitor Vincenzo Maculani to be charged. Throughout his trial Galileo steadfastly maintained that since 1616 he had faithfully kept his promise not to hold any of the condemned opinions, and initially he denied even defending them. However, he was eventually persuaded to admit that, contrary to his true intention, a reader of his Dialogue could well have obtained the impression that it was intended to be a defence of Copernicanism. In view of Galileo's rather implausible denial that he had ever held Copernican ideas after 1616 or ever intended to defend them in the Dialogue, his final interrogation, in July 1633, concluded with his being threatened with torture if he did not tell the truth, but he maintained his denial despite the threat.

The sentence of the Inquisition was delivered on 22 June. It was in three essential parts:

Galileo was found "vehemently suspect of heresy", namely of having held the opinions that the Sun lies motionless at the centre of the universe, that the Earth is not at its centre and moves, and that one may hold and defend an opinion as probable after it has been declared contrary to Holy Scripture. He was required to "abjure, curse and detest" those opinions.

He was sentenced to formal imprisonment at the pleasure of the Inquisition. On the following day this was commuted to house arrest, which he remained under for the rest of his life.

His offending Dialogue was banned; and in an action not announced at the trial, publication of any of his works was forbidden, including any he might write in the future.

According to popular legend, after recanting his theory that the Earth moved around the Sun, Galileo allegedly muttered the rebellious phrase And yet it moves. A 1640s painting by the Spanish painter Bartolomé Esteban Murillo or an artist of his school, in which the words were hidden until restoration work in 1911, depicts an imprisoned Galileo apparently gazing at the words "E pur si muove" written on the wall of his dungeon. The earliest known written account of the legend dates to a century after his death, but Stillman Drake writes "there is no doubt now that the famous words were already attributed to Galileo before his death".

After a period with the friendly Ascanio Piccolomini (the Archbishop of Siena), Galileo was allowed to return to his villa at Arcetri near Florence in 1634, where he spent the remainder of his life under house arrest. Galileo was ordered to read the seven penitential psalms once a week for the next three years. However, his daughter Maria Celeste relieved him of the burden after securing ecclesiastical permission to take it upon herself.

It was while Galileo was under house arrest that he dedicated his time to one of his finest works, Two New Sciences. Here he summarised work he had done some forty years earlier, on the two sciences now called kinematics and strength of materials, published in Holland to avoid the censor. This book has received high praise from Albert Einstein. As a result of this work, Galileo is often called the "father of modern physics". He went completely blind in 1638 and was suffering from a painful hernia and insomnia, so he was permitted to travel to Florence for medical advice.

Dava Sobel argues that prior to Galileo's 1633 trial and judgement for heresy, Pope Urban VIII had become preoccupied with court intrigue and problems of state, and began to fear persecution or threats to his own life. In this context, Sobel argues that the problem of Galileo was presented to the pope by court insiders and enemies of Galileo. Having been accused of weakness in defending the church, Urban reacted against Galileo out of anger and fear.

Death

Galileo continued to receive visitors until 1642, when, after suffering fever and heart palpitations, he died on 8 January 1642, aged 77. The Grand Duke of Tuscany, Ferdinando II, wished to bury him in the main body of the Basilica of Santa Croce, next to the tombs of his father and other ancestors, and to erect a marble mausoleum in his honour.

These plans were dropped, however, after Pope Urban VIII and his nephew, Cardinal Francesco Barberini, protested, because Galileo had been condemned by the Catholic Church for "vehement suspicion of heresy". He was instead buried in a small room next to the novices' chapel at the end of a corridor from the southern transept of the basilica to the sacristy. He was reburied in the main body of the basilica in 1737 after a monument had been erected there in his honour; during this move, three fingers and a tooth were removed from his remains.

Tomb of Galileo, Santa Croce, Florence

One of these fingers, the middle finger from Galileo's right hand, is currently on exhibition at the Museo Galileo in Florence, Italy.

Scientific Contributions

Scientific Methods

Galileo made original contributions to the science of motion through an innovative combination of experiment and mathematics. More typical of science at the time were the qualitative studies of William Gilbert, on magnetism and electricity. Galileo's father, Vincenzo Galilei, a lutenist and music theorist, had performed experiments establishing perhaps the oldest known non-linear relation in physics: for a stretched string, the pitch varies as the square root of the tension. These observations lay within the framework of the Pythagorean tradition of music, well-known to instrument makers, which included the fact that subdividing a string by a whole number produces a harmonious scale. Thus, a limited amount of mathematics had long related music and physical science, and young Galileo could see his own father's observations expand on that tradition.

- ROBERT STAWELL BALL

Galileo was one of the first modern thinkers to clearly state that the laws of nature are mathematical. In *The Assayer* he wrote "Philosophy is written in this grand book, the universe ... It is written in the language of mathematics, and its characters are triangles, circles, and other geometric figures;...." His mathematical analyses are a further development of a tradition employed by late scholastic natural philosophers, which Galileo learned when he studied philosophy. His work marked another step towards the eventual separation of science from both philosophy and religion; a major development in human thought. He was often willing to change his views in accordance with observation. In order to perform his experiments, Galileo had to set up standards of length and time, so that measurements made on different days and in different laboratories could be compared in a reproducible fashion. This provided a reliable foundation on which to confirm mathematical laws using inductive reasoning.

Galileo showed a modern appreciation for the proper relationship between mathematics, theoretical physics, and experimental physics. He understood the parabola, both in terms of conic sections and in terms of the ordinate (y) varying as the square of the abscissa (x). Galilei further asserted that the parabola was the theoretically ideal trajectory of a uniformly accelerated projectile in the absence of air resistance or other disturbances. He conceded that there are limits to the validity of this theory, noting on theoretical grounds that a projectile trajectory of a size comparable to that of the Earth could not possibly be a parabola, but he nevertheless maintained that for distances up to the range of the artillery of his day, the deviation of a projectile's trajectory from a parabola would be only very slight.

Astronomy

Galileo showed the Doge of Venice how to use the telescope (Fresco by Giuseppe Bertini)

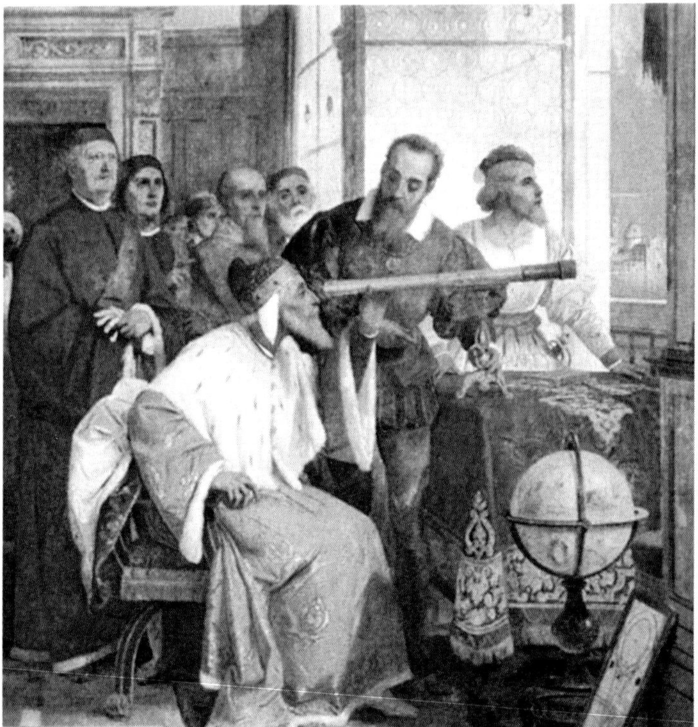

Galileo showed the Doge of Venice how to use the telescope (Fresco by Giuseppe Bertini)

It was on this page that Galileo first noted an observation of the moons of Jupiter. This observation upset the notion that all celestial bodies must revolve around the Earth. Galileo published a full description in Sidereus Nuncius in March 1610. The phases of Venus, observed by Galileo in 1610.

Based only on uncertain descriptions of the first practical telescope which Hans Lippershey tried to patent in the Netherlands in 1608, Galileo, in the following year, made a telescope with about 3x magnification. He later made improved versions with up to about 30x magnification. With a Galilean telescope, the observer could see magnified, upright images on the earth—it was what is commonly known as a terrestrial telescope or a spyglass. He could also use it to observe the sky; for a time he was one of those who could construct telescopes good enough for that purpose. On 25 August 1609, he demonstrated one of his early telescopes, with a magnification of about 8 or 9, to Venetian lawmakers. His telescopes were also a profitable sideline for Galileo, who sold them to merchants who found them useful both at sea and as items of trade. He published his initial telescopic astronomical observations in March 1610 in a brief treatise entitled Sidereus Nuncius (Starry Messenger).

Kepler's Supernova

Tycho and others had observed the supernova of 1572. Ottavio Brenzoni's letter of 15 January 1605 to Galileo brought the 1572 supernova and the less bright nova of 1601 to Galileo's notice. Galileo observed and discussed Kepler's supernova in 1604. Since these new stars displayed no detectable diurnal parallax, Galileo concluded that they were distant stars, and therefore disproved the Aristotelian belief in the immutability of the heavens.

Jupiter's Moons

On 7 January 1610, Galileo observed with his telescope what he described at the time as "three fixed stars, totally invisible by their smallness", all close to Jupiter, and lying on a straight line through it. Observations on subsequent nights showed that the positions of these "stars" relative to Jupiter were changing in a way that would have been inexplicable if they had really been fixed stars. On 10 January, Galileo noted that one of them had disappeared, an observation which he attributed to its being hidden behind Jupiter. Within a few days, he concluded that they were orbiting Jupiter: he had discovered three of Jupiter's four largest moons. He discovered the fourth on 13 January. Galileo named the group of four the Medicean stars, in honour of his future patron, Cosimo II de' Medici, Grand Duke of Tuscany, and Cosimo's three brothers. Later astronomers, however, renamed them *Galilean satellites* in honour of their discoverer. These satellites are now called Io, Europa, Ganymede, and Callisto.

- ROBERT STAWELL BALL

His observations of the satellites of Jupiter caused a revolution in astronomy: a planet with smaller planets orbiting it did not conform to the principles of Aristotelian cosmology, which held that all heavenly bodies should circle the Earth, and many astronomers and philosophers initially refused to believe that Galileo could have discovered such a thing. His observations were confirmed by the observatory of Christopher Clavius and he received a hero's welcome when he visited Rome in 1611. Galileo continued to observe the satellites over the next eighteen months, and by mid-1611, he had obtained remarkably accurate estimates for their periods—a feat which Kepler had believed impossible.

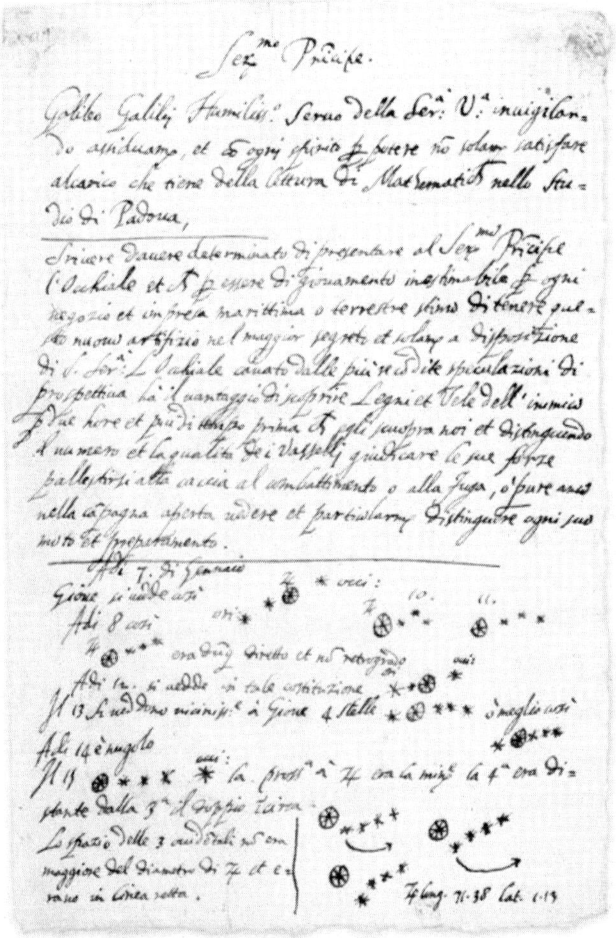

It was on this page that Galileo first noted an observation of the moons of Jupiter. This observation upset the notion that all celestial bodies must revolve around the Earth. Galileo

---

- ROBERT STAWELL BALL

published a full description in Sidereus Nuncius in March 1610.

Venus, Saturn, and Neptune

From September 1610, Galileo observed that Venus exhibited a full set of phases similar to that of the Moon. The heliocentric model of the solar system developed by Nicolaus Copernicus predicted that all phases would be visible since the orbit of Venus around the Sun would cause its illuminated hemisphere to face the Earth when it was on the opposite side of the Sun and to face away from the Earth when it was on the Earth-side of the Sun. On the other hand, in Ptolemy's geocentric model it was impossible for any of the planets' orbits to intersect the spherical shell carrying the Sun. Traditionally the orbit of Venus was placed entirely on the near side of the Sun, where it could exhibit only crescent and new phases. It was, however, also possible to place it entirely on the far side of the Sun, where it could exhibit only gibbous and full phases. After Galileo's telescopic observations of the crescent, gibbous and full phases of Venus, therefore, this

Ptolemaic model became untenable. Thus in the early 17th century as a result of his discovery the great majority of astronomers converted to one of the various geo-heliocentric planetary models, such as the Tychonic, Capellan and Extended Capellan models, each either with or without a daily rotating Earth. These all had the virtue of explaining the phases of Venus without the vice of the 'refutation' of full heliocentrism's prediction of stellar parallax. Galileo's discovery of the phases of Venus was thus arguably his most empirically practically influential contribution to the two-stage transition from full geocentrism to full heliocentrism via geo-heliocentrism.

Galileo observed the planet Saturn, and at first mistook its rings for planets, thinking it was a three-bodied system. When he observed the planet later, Saturn's rings were directly oriented at Earth, causing him to think that two of the bodies had disappeared. The rings reappeared when he observed the planet in 1616, further confusing him.

Galileo also observed the planet Neptune in 1612. It appears in his notebooks as one of many unremarkable dim stars. He did not realise that it was a planet, but he did note its motion relative to the stars before losing track of it.

Galileo also observed the planet Neptune in 1612. It appears in his notebooks as one of many unremarkable dim stars. He did not realise that it was a planet, but he did note its motion relative to the stars before losing track of it.

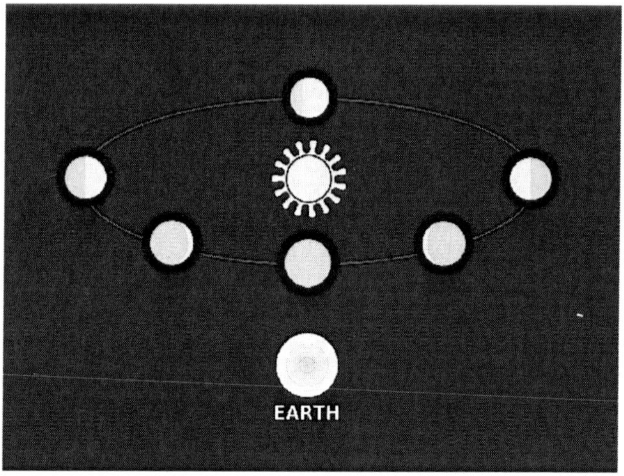

The phases of Venus, observed by Galileo in 1610

Sunspots

Galileo made naked-eye and telescopic studies of sunspots. Their existence raised another difficulty with the unchanging perfection of the heavens as posited in orthodox Aristotelian celestial physics. An apparent annual variation in their trajectories, observed by Francesco Sizzi and others in 1612–1613, also provided a powerful argument against both the Ptolemaic system and the geoheliocentric system of Tycho Brahe. A dispute over claimed priority in the discovery of sunspots, and in their interpretation, led Galileo to a long and bitter feud with the Jesuit Christoph Scheiner. In the middle was Mark Welser, to whom Scheiner had announced his discovery, and who asked Galileo for his opinion. In fact, there is little doubt that both of them were beaten by David Fabricius and his son Johannes.

- ROBERT STAWELL BALL

Moon

Prior to Galileo's construction of his version of a telescope, Thomas Harriot, an English mathematician and explorer, had already used what he dubbed a "perspective tube" to observe the moon. Reporting his observations, Harriot noted only "strange spottednesse" in the waning of the crescent, but was ignorant to the cause. Galileo, due in part to his artistic training and the knowledge of chiaroscuro, had understood the patterns of light and shadow were in fact topographical markers. While not being the only one to observe the moon through a telescope, Galileo was the first to deduce the cause of the uneven waning as light occlusion from lunar mountains and craters. In his study he also made topographical charts, estimating the heights of the mountains. The moon was not what was long thought to have been a translucent and perfect sphere, as Aristotle claimed, and hardly the first "planet", an "eternal pearl to magnificently ascend into the heavenly empyrian", as put forth by Dante.

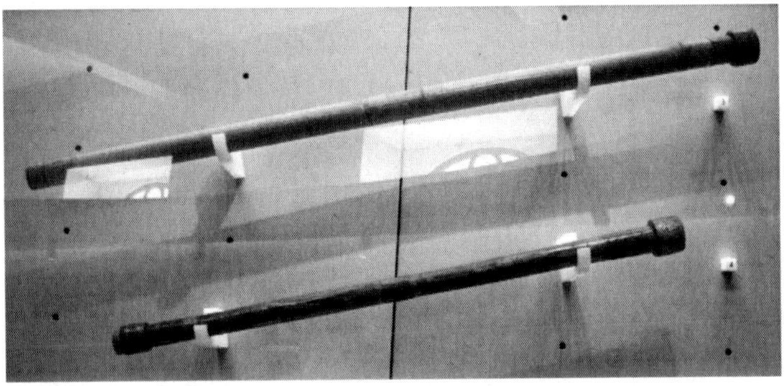

Galileo's "cannocchiali" telescopes at the Museo Galileo, Florence

Milky Way and Stars

Galileo observed the Milky Way, previously believed to be nebulous, and found it to be a multitude of stars packed so densely that they appeared from Earth to be clouds. He located many other stars too distant to be visible with the naked eye. He observed the double star Mizar in Ursa Major in 1617.

In the Starry Messenger, Galileo reported that stars appeared as mere blazes of light, essentially unaltered in appearance by the telescope, and contrasted them to planets, which the telescope revealed to be discs. But shortly thereafter, in his letters on sunspots, he reported that the telescope revealed the shapes of both stars and planets to be "quite round". From that point forward, he continued to report that telescopes showed the roundness of stars, and that stars seen through the telescope measured a few seconds of arc in diameter. He also devised a method for measuring the apparent size of a star without a telescope. As described in his Dialogue Concerning the two Chief World Systems, his method was to hang a thin rope in his line of sight to the star and measure the maximum distance from which it would wholly obscure the star. From his measurements of this distance and of the width of the rope, he could calculate the angle subtended by the star at his viewing point. In his Dialogue, he reported that he had found the apparent diameter of a star of first magnitude to be no more than 5 arcseconds, and that of one of sixth magnitude to be about 5/6 arcseconds. Like most astronomers of his day, Galileo did not recognise that the apparent sizes of stars that he measured were spurious, caused by diffraction and atmospheric distortion (see seeing disk or Airy disk), and did not represent the true sizes of stars. However, Galileo's values were much smaller than previous estimates of the apparent sizes of the brightest stars, such as those made by Tycho Brahe (see Magnitude) and enabled Galileo to counter anti-Copernican arguments such as those made by Tycho that these stars would have to be absurdly large for their annual parallaxes to be undetectable. Other astronomers such as Simon Marius, Giovanni Battista Riccioli, and Martinus Hortensius made similar measurements of stars, and Marius and Riccioli concluded the smaller sizes were not small enough to answer Tycho's argument.

Engineering

Galileo made a number of contributions to what is now known as engineering, as distinct from pure physics. Between 1595 and 1598, Galileo devised and improved a Geometric and military compass suitable for use by gunners and surveyors. This expanded on earlier instruments designed by Niccolò Tartaglia and Guidobaldo del Monte. For gunners, it offered, in addition to a new and safer way of elevating cannons accurately, a way of quickly computing the charge of gunpowder for cannonballs of different sizes and materials. As a geometric instrument, it enabled the construction of any regular polygon, computation of the area of any polygon or circular sector, and a variety of other calculations. Under Galileo's direction, instrument maker Marc'Antonio Mazzoleni produced more than 100 of these compasses, which Galileo sold (along with an instruction manual he wrote) for 50 lire and offered a course of instruction in the use of the compasses for 120 lire.

Galileo's geometrical and military compass, thought to have been made c. 1604 by his personal instrument-maker Marc'Antonio Mazzoleni

In about 1593, Galileo constructed a thermometer, using the expansion and contraction of air in a bulb to move water in an attached tube.

In 1609, Galileo was, along with Englishman Thomas Harriot and others, among the first to use a refracting telescope as an instrument to observe stars, planets or moons. The name "telescope" was coined for Galileo's instrument by a Greek mathematician,

Giovanni Demisiani, at a banquet held in 1611 by Prince Federico Cesi to make Galileo a member of his Accademia dei Lincei. The name was derived from the Greek tele = 'far' and skopein = 'to look or see'. In 1610, he used a telescope at close range to magnify the parts of insects. By 1624 Galileo had used a compound microscope. He gave one of these instruments to Cardinal Zollern in May of that year for presentation to the Duke of Bavaria, and in September he sent another to Prince Cesi. The Linceans played a role again in naming the "microscope" a year later when fellow academy member Giovanni Faber coined the word for Galileo's invention from the Greek words μικρόν (micron) meaning "small", and σκοπεῖν (skopein) meaning "to look at". The word was meant to be analogous with "telescope". Illustrations of insects made using one of Galileo's microscopes, and published in 1625, appear to have been the first clear documentation of the use of a compound microscope.

In 1612, having determined the orbital periods of Jupiter's satellites, Galileo proposed that with sufficiently accurate knowledge of their orbits, one could use their positions as a universal clock, and this would make possible the determination of longitude. He worked on this problem from time to time during the remainder of his life; but the practical problems were severe. The method was first successfully applied by Giovanni Domenico Cassini in 1681 and was later used extensively for large land surveys; this method, for example, was used to survey France, and later by Zebulon Pike of the midwestern United States in 1806. For sea navigation, where delicate telescopic observations were more difficult, the longitude problem eventually required development of a practical portable marine chronometer, such as that of John Harrison. Late in his life, when totally blind, Galileo designed an escapement mechanism for a pendulum clock (called Galileo's escapement), although no clock using this was built until after the first fully operational pendulum clock was made by Christiaan Huygens in the 1650s.

- ROBERT STAWELL BALL

Physics

Galileo's theoretical and experimental work on the motions of bodies, along with the largely independent work of Kepler and René Descartes, was a precursor of the classical mechanics developed by Sir Isaac Newton. Galileo conducted several experiments with pendulums. It is popularly believed (thanks to the biography by Vincenzo Viviani) that these began by watching the swings of the bronze chandelier in the cathedral of Pisa, using his pulse as a timer. Later experiments are described in his *Two New Sciences*. Galileo claimed that a simple pendulum is isochronous, i.e. that its swings always take the same amount of time, independently of the amplitude. In fact, this is only approximately true, as was discovered by Christiaan Huygens. Galileo also found that the square of the period varies directly with the length of the pendulum. Galileo's son, Vincenzo, sketched a clock based on his father's theories in 1642. The clock was never built and, because of the large swings required by its verge escapement, would have been a poor timekeeper. (See Engineering above.)

Galileo e *Viviani*, 1892, Tito Lessi

Galileo is lesser known for, yet still credited with, being one of the first to understand sound frequency. By scraping a chisel at different speeds, he linked the pitch of the sound produced to the spacing of the chisel's skips, a measure of frequency. In 1638, Galileo described an experimental method to measure the speed of light by arranging that two observers, each having lanterns equipped with shutters, observe each other's lanterns at some distance. The first observer opens the shutter of his lamp, and, the second, upon seeing the light, immediately opens the shutter of his own lantern. The time between the first observer's opening his shutter and seeing the light from the second observer's lamp indicates the time it takes light to travel back and forth between the two observers. Galileo reported that when he tried this at a distance of less than a mile, he was unable to determine whether or not the light appeared instantaneously. Sometime between Galileo's death and 1667, the members of the Florentine *Accademia del Cimento* repeated the experiment over a distance of about a mile and obtained a similarly inconclusive result. We now know that the speed of light is far too fast to be measured by such methods (with human shutter-openers on Earth).

Galileo put forward the basic principle of relativity, that the laws of physics are the same in any system that is moving at a constant speed in a straight line, regardless of its particular speed or direction. Hence, there is no absolute motion or absolute rest. This principle provided the basic framework for Newton's laws of motion and is central to Einstein's special theory of relativity.

Falling Bodies

A biography by Galileo's pupil Vincenzo Viviani stated that Galileo had dropped balls of the same material, but different masses, from the Leaning Tower of Pisa to demonstrate that their time of descent was independent of their mass. This was contrary to what Aristotle had taught: that heavy objects fall faster than lighter ones, in direct proportion to weight. While this story has been retold in popular accounts, there is no account by Galileo himself of such an experiment, and it is generally accepted by historians that it was at most a thought experiment which did not actually take place. An exception is Drake, who argues that the experiment did take place, more or less as Viviani described it. The experiment described was actually performed by Simon Stevin (commonly known as Stevinus) and Jan Cornets de Groot, although the building used was actually the church tower in Delft in 1586. However most of his experiments with falling bodies were carried out using inclined planes where both the issues of timing and air resistance were much reduced.

In his 1638 Discorsi, Galileo's character Salviati, widely regarded as Galileo's spokesman, held that all unequal weights would fall with the same finite speed in a vacuum. But this had previously been proposed by Lucretius and Simon Stevin. Cristiano Banti's Salviati also held it could be experimentally demonstrated by the comparison of pendulum motions in air with bobs of lead and of cork which had different weight but which were otherwise similar.

Galileo proposed that a falling body would fall with a uniform acceleration, as long as the resistance of the medium through which it was falling remained negligible, or in the limiting case of its falling through a vacuum. He also derived the correct kinematical law for the distance travelled during a uniform acceleration starting from rest—namely, that it is proportional to the square of the elapsed time ($d \propto t 2$). Prior to Galileo, Nicole Oresme, in the 14th century, had derived the times-squared law for uniformly accelerated change, and Domingo de Soto had suggested in the 16th century that bodies falling through a homogeneous medium would be uniformly accelerated. Galileo expressed the time-squared law using geometrical constructions and mathematically precise words, adhering to the standards of the day. (It remained for others to re-express the law in algebraic terms).

He also concluded that objects retain their velocity in the absence of any impediments to their motion, thereby contradicting the generally accepted Aristotelian hypothesis that a body could only remain in so-called "violent", "unnatural", or "forced" motion so long as an agent of change (the "mover") continued to act on it. Philosophical ideas relating to inertia had been proposed by John Philoponus and Jean Buridan. Galileo stated: "Imagine any particle projected along a horizontal plane without friction; then we know, from what has been more fully explained in the preceding pages, that this particle will move along this same plane with a motion which is uniform and perpetual, provided the plane has no limits" This was incorporated into Newton's laws of motion (first law).

Mathematics

While Galileo's application of mathematics to experimental physics was innovative, his mathematical methods were the standard ones of the day, including dozens of examples of an inverse proportion square root method passed down from Fibonacci and Archimedes. The analysis and proofs relied heavily on the Eudoxian theory of proportion, as set forth in the fifth book of Euclid's Elements.

This theory had become available only a century before, thanks to accurate translations by Tartaglia and others; but by the end of Galileo's life, it was being superseded by the algebraic methods of Descartes.

The concept now named Galileo's paradox was not original with him. His proposed solution, that infinite numbers cannot be compared, is no longer considered useful.

Dome of the Cathedral of Pisa with the "lamp of Galileo"

Dome depicting the Ascension of the Blessed Virgin by Riminaldi; cathedral, Pisa, Italy Equally seen is "the lamp of Galileo" (leading him to invent the law of the isochrony of the pendulum). - work by Vincenzo Possenti in 1586.

Timeline

- 1543 – Nicolaus Copernicus publishes *De revolutionibus orbium coelestium* as an alternative world system to the Ptolemy's geocentric model causing subsequent questions to be raised about Aristotelian physics following Copernicus' death

- 1563 – Parents Vincenzo Galilei and Giulia Ammannati marry

- **1564 – Birth** in Pisa, Italy

- ~1570 – Thomas Digges publishes *Pantometria* describing a telescope built between 1540–1559 by his father Leonard Digges

- 1573 – Tycho Brahe publishes *De nova stella* (*On the new star*) refuting Aristotelian belief in immutable celestial spheres and an eternal, unchanging, more perfect heavenly realm of celestial aether above the moon

- 1576 – Giuseppe Moletti, Galileo's predecessor in the mathematics chair at Padua, reports falling bodies of the same shape fall at the same speed, regardless of material

- 1581 – His father, Vincenzo Galilei publishes *Dialogo della musica antica et moderna* formulating musical theories

- 1581 – Enrols as medical student at University of Pisa

- 1582 – Attends mathematics lecture by Ostilio Ricci and decides to study math and science

- 1585 – Leaves University of Pisa without degree and works as tutor

- 1586 – Invents hydrostatic balance; wrote *La Balancitta* (*The little balance*)

- 1586 – Simon Stevin publishes results for dropping lead weights from 10 meters

- 1588 – Tycho Brahe publishes work on comets containing a description of the Tychonic system of the world

- 1589 – Appointed to Mathematics Chair, University of Pisa

- ~1590 – Partially completes *De Motu* (*On Motion*), which is never published

- 1591 – Death of his father, Vicenzo Galilei

- 1592 – Appointed professor of mathematics at University of Padua, remains 18 years

- ~1593 – Invents early thermometer that unfortunately depended on both temperature and pressure

- ~1595 – Invents improved ballistics calculation geometric and military compass, which he later improves for surveying and general calculations and earns income from tutoring on its use

- 1597 – Letter to Kepler indicates his belief in the Copernican System

- 1600 – First child, Virginia is born; ~1600 *Le Meccaniche* (*Mechanics*)

- 1600 – William Gilbert publishes *On the Magnet and Magnetic Bodies, and on That Great Magnet the Earth* with arguments supporting the Copernican system

- 1600 – Roman Inquisition finds Giordano Bruno, Copernican system supporter, guilty of heresy for opinions on pantheism and the eternal plurality of worlds, and for denial of the Trinity, divinity of Christ, virginity of Mary, and Transubstantiation; burned at the stake by civil authorities

- 1601 – Daughter Livia is born

- 1604 – Measures supernova position indicating no parallax for the new star

- 1605 – Sued by brothers-in-law for nonpayment of sisters' dowries

- 1606 – Son Vincenzo born

- 1606 – Publishes manual for his calculating compass

- 1607 – Rotilio Orlandini attempts to assassinate Galileo's friend, Friar Paolo Sarpi

- 1608 – Hans Lippershey invents a refracting telescope

- 1609 – Independently invents and improves telescopes based on description of invention by Hans Lippershey

- 1609 – Kepler publishes *Astronomia nova* containing his first two laws and for the first time demonstrates the Copernican model is more accurate than the Ptolemaic for uses such as navigation and prediction

- 1609 – Thomas Harriot sketches the Moon from telescopic observations made four months before Galileo's

- 1610 – Publishes *Sidereus Nuncius* (*Starry Messenger*); views our moon's mountains and craters and brightest 4 of Jupiter's moons

- 1610 – Martin Horky publishes Brevissima Peregrinatio Contra Nuncium Sidereum, opposing Galileo

- 1610 – Kepler requests one of Galileo's telescopes or lenses, but Galileo replies he is too busy to build one and has no extras

- 1610 – Lifetime appointment to mathematics position at University of Padua, and as mathematician and philosopher for Cosimo II, Grand

- Duke of Tuscany

- 1611 – Discovers phases of Venus; granted audience with Pope; made member of Lincean Academy

- 1611 – Awarded an honorary degree by the Jesuit College in Rome

- 1611 – David Fabricius publishes *Narration on Spots Observed on the Sun and their Apparent Rotation with the Sun* prior to Christoph Scheiner and Galileo's published works on the subject

- 1612 – Proposed Jupiter's moons could be used as a universal clock for possible determination of longitude

- ~1612 or 1613 – Francesco Sizzi discovers annual variations in sunspots' motions

- 1613 – *Letters on Sunspots*

- 1613 – Letter to Benedetto Castelli discussing the rotation of the sun and Galileo's support of the Copernican system. Using Biblical inerrancy as a basis, Galileo writes that Joshua's command for the sun to stand still in Joshua 10:13 proves the "impossibility of the Aristotelian and Ptolemaic world system, and on the other hand agrees very well with the Copernican one." He went on to write that "the sun gives not only light (as it obviously does) but also motion to all the planets that revolve around it" by its rotation.

- 1615 – *Letter to Grand Duchess Christina* (not published until 1636)

- 1616 – Officially warned by the Church not to hold or defend the Copernican System

- 1616 – The Catholic Church places *De revolutionibus orbium coelestium* on the List of Prohibited Books, pending correction

- 1616 – Private letter "Discourse on the Tides"

- 1617 – Moves into Bellosguardo, west of Florence, near his daughters' convent; observes double star Mizar in Ursa Major

- 1619 – Kepler publishes *Harmonices Mundi* which introduces his third law

- 1619 – *Discourse on the Comets*

- 1623 – Maffeo Barberini becomes Pope Urban VIII

- 1623 – Publishes *The Assayer*

- 1624 – Visits Pope who praises and honours him, leaving with assumed permission to publish work on the Copernican vs. Ptolemaic Systems; used a compound microscope

- 1625 – Illustrations of insects made using one of Galileo's microscopes published

- 1630 – Completes *Dialogue Concerning the Two Chief World Systems* and subsequently receives approval of Church censor

- 1632 – Publishes *Dialogue Concerning the Two Chief World Systems*

- 1633 – sentenced by the Inquisition to imprisonment, commuted to house arrest, for vehement suspicion of heresy in violating the 1616 injunction

- 1633 – Catholic Church places *Dialogue Concerning the Two Chief World Systems* on the List of Prohibited Books

- 1638 – Publishes *Dialogues Concerning Two New Sciences*

- 1642 – Death in Arcetri, Italy

- 1668 – Isaac Newton builds his reflecting telescope

- 1687 – Newton publishes *Philosophiæ Naturalis Principia Mathematica* deriving Kepler's laws from the Universal Law of Gravitation and the Laws of Motion

- 2016 – The American *Juno* spacecraft, carrying a plaque and a Lego minifigure dedicated at Galileo, arrives at Jupiter's orbit

Later Church Reassessments

The Galileo affair was largely forgotten after Galileo's death, and the controversy subsided. The Inquisition's ban on reprinting Galileo's works was lifted in 1718 when permission was granted to publish an edition of his works (excluding the condemned Dialogue) in Florence. In 1741 Pope Benedict XIV authorised the publication of an edition of Galileo's complete scientific works which included a mildly censored version of the Dialogue. In 1758 the general prohibition against works advocating heliocentrism was removed from the Index of prohibited books, although the specific ban on uncensored versions of the Dialogue and Copernicus's De Revolutionibus remained. All traces of official opposition to heliocentrism by the church disappeared in 1835 when these works were finally dropped from the Index.

Interest in the Galileo affair was revived in the early 19th century, when Protestant polemicists used it (and other events such as the Spanish Inquisition and the myth of the flat Earth) to attack Roman Catholicism. Interest in it has waxed and waned ever since. In 1939 Pope Pius XII, in his first speech to the Pontifical Academy of Sciences, within a few months of his election to the papacy, described Galileo as being among the "most audacious heroes of research... not afraid of the stumbling blocks and the risks on the way, nor fearful of the funereal monuments". His close advisor of 40 years, Professor Robert Leiber, wrote: "Pius XII was very careful not to close any doors (to science) prematurely. He was energetic on this point and regretted that in the case of Galileo."

On 15 February 1990, in a speech delivered at the Sapienza University of Rome, Cardinal Ratzinger (later to become Pope Benedict XVI) cited some current views on the Galileo affair as forming what he called "a symptomatic case that permits us to see how deep the self-doubt of the modern age, of science and technology goes today". Some of the views he cited were those of the philosopher Paul Feyerabend, whom he quoted as saying "The Church at the time of Galileo kept much more closely to reason than did Galileo himself, and she took into consideration the ethical and social consequences of Galileo's teaching too. Her verdict against Galileo was rational and just and the revision of this verdict can be justified only on the grounds of what is politically opportune." The Cardinal did not clearly indicate whether he agreed or disagreed with Feyerabend's assertions. He did, however, say "It would be foolish to construct an impulsive apologetic on the basis of such views."

On 31 October 1992, Pope John Paul II expressed regret for how the Galileo affair was handled, and issued a declaration acknowledging the errors committed by the Catholic Church tribunal that judged the scientific positions of Galileo Galilei, as the result of a study conducted by the Pontifical Council for Culture. In March 2008 the head of the Pontifical Academy of Sciences, Nicola Cabibbo, announced a plan to honour Galileo by erecting a statue of him inside the Vatican walls. In December of the same year, during events to mark the 400th anniversary of Galileo's earliest telescopic observations, Pope Benedict XVI praised his contributions to astronomy. A month later, however, the head of the Pontifical Council for Culture, Gianfranco Ravasi, revealed that the plan to erect a statue of Galileo in the grounds of the Vatican had been suspended.

Impact on modern science

According to Stephen Hawking, Galileo probably bears more of the responsibility for the birth of modern science than anybody else, and Albert Einstein called him the father of modern science.

Galileo's astronomical discoveries and investigations into the Copernican theory have led to a lasting legacy which includes the categorisation of the four large moons of Jupiter discovered by Galileo (Io, Europa, Ganymede and Callisto) as the Galilean moons. Other scientific endeavours and principles are named after Galileo including the Galileo spacecraft, the first spacecraft to enter orbit around Jupiter, the proposed Galileo global satellite navigation system, the transformation between inertial systems in classical mechanics denoted Galilean transformation and the Gal (unit), sometimes known as the Galileo, which is a non-SI unit of acceleration.

Partly because 2009 was the fourth centenary of Galileo's first recorded astronomical observations with the telescope, the United Nations scheduled it to be the International Year of Astronomy. A global scheme was laid out by the International Astronomical Union (IAU), also endorsed by UNESCO—the UN body responsible for educational, scientific and cultural matters. The International Year of Astronomy 2009 was intended to be a global celebration of astronomy and its contributions to society and culture, stimulating worldwide interest not only in astronomy but science in general, with a particular slant towards young people.

Asteroid 697 Galilea is named in his honour.

In Artistic and Popular Media

Galileo is mentioned several times in the "opera" section of the Queen song, "Bohemian Rhapsody". He features prominently in the song "Galileo" performed by the Indigo Girls and Amy Grant's Galileo on her Heart in Motion album.

Twentieth-century plays have been written on Galileo's life, including *Life of Galileo* (1943) by the German playwright Bertolt Brecht, with a film adaptation (1975) of it, and *Lamp At Midnight* (1947) by Barrie Stavis, as well as the 2008 play "Galileo Galilei".

Kim Stanley Robinson wrote a science fiction novel entitled *Galileo's Dream* (2009), in which Galileo is brought into the future to help resolve a crisis of scientific philosophy; the story moves back and forth between Galileo's own time and a hypothetical distant future, and contains a great deal of biographical information.

Galileo Galilei was recently selected as a main motif for a high value collectors' coin: the €25 International Year of Astronomy commemorative coin, minted in 2009. This coin also commemorates the 400th anniversary of the invention of Galileo's telescope. The obverse shows a portion of his portrait and his telescope. The background shows one of his first drawings of the surface of the moon. In the silver ring other telescopes are depicted: the Isaac Newton Telescope, the observatory in Kremsmünster Abbey, a modern telescope, a radio telescope and a space telescope. In 2009, the Galileoscope was also released. This is a mass-produced, low-cost educational 2-inch (51 mm) telescope with relatively high quality.

Writings

Galileo's early works describing scientific instruments include the 1586 tract entitled The Little Balance (La Billancetta) describing an accurate balance to weigh objects in air or water and the 1606 printed manual Le Operazioni del Compasso Geometrico et Militare on the operation of a geometrical and military compass.

Statue outside the Uffizi, Florence

His early works in dynamics, the science of motion and mechanics were his circa 1590 Pisan De Motu (On Motion) and his circa 1600 Paduan Le Meccaniche (Mechanics). The former was based on Aristotelian–Archimedean fluid dynamics and held that the speed of gravitational fall in a fluid medium was proportional to the excess of a body's specific weight over that of the medium, whereby in a vacuum, bodies would fall with speeds in proportion to their specific weights. It also subscribed to the Philoponan impetus dynamics in which impetus is self-dissipating and free-fall in a vacuum would have an essential terminal speed according to specific weight after an initial period of acceleration.

Galileo's 1610 *The Starry Messenger* (*Sidereus Nuncius*) was the first scientific treatise to be published based on observations made through a telescope. It reported his discoveries of:

- the Galilean moons
- the roughness of the Moon's surface
- the existence of a large number of stars invisible to the naked eye,

particularly those responsible for the appearance of the Milky Way

- differences between the appearances of the planets and those of the fixed stars—the former appearing as small discs, while the latter appeared as unmagnified points of light

Galileo published a description of sunspots in 1613 entitled *Letters on Sunspots* suggesting the Sun and heavens are corruptible. The *Letters on Sunspots* also reported his 1610 telescopic observations of the full set of phases of Venus, and his discovery of the puzzling "appendages" of Saturn and their even more puzzling subsequent disappearance. In 1615 Galileo prepared a manuscript known as the "Letter to the Grand Duchess Christina" which was not published in printed form until 1636. This letter was a revised version of the Letter to Castelli, which was denounced by the Inquisition as an incursion upon theology by advocating Copernicanism both as physically true and as consistent with Scripture. In 1616, after the order by the inquisition for Galileo not to hold or defend the Copernican position, Galileo wrote the "Discourse on the Tides" (Discorso sul flusso e il reflusso del mare) based on the Copernican earth, in the form of a private letter to Cardinal Orsini. In 1619, Mario Guiducci, a pupil of Galileo's, published a lecture written largely by Galileo under the title Discourse on the Comets (Discorso Delle Comete), arguing against the Jesuit interpretation of comets.

In 1623, Galileo published The Assayer—Il Saggiatore, which attacked theories based on Aristotle's authority and promoted experimentation and the mathematical formulation of scientific ideas. The book was highly successful and even found support among the higher echelons of the Christian church. Following the success of The Assayer, Galileo published the *Dialogue Concerning the Two Chief World Systems* (Dialogo sopra i due massimi sistemi del mondo) in 1632. Despite taking care to adhere to the Inquisition's 1616 instructions, the claims in the book favouring Copernican theory and a non Geocentric model of the solar system led to

Galileo being tried and banned on publication. Despite the publication ban, Galileo published his *Discourses and Mathematical Demonstrations Relating to Two New Sciences (Discorsi e Dimostrazioni Matematiche, intorno a due nuove scienze)* in 1638 in Holland, outside the jurisdiction of the Inquisition.

Published Written Works

Galileo's main written works are as follows:

- *The Little Balance* (1586; in Italian: *La Billancetta*)
- *On Motion* (c. 1590; in Latin: *De Motu Antiquiora*)
- *Mechanics* (c. 1600; in Italian: *Le mecaniche*)
- *The Operations of Geometrical and Military Compass* (1606; in Italian: *Le operazioni del compasso geometrico et militare*)
- *The Starry Messenger* (1610; in Latin: *Sidereus Nuncius*)
- *Discourse on Floating Bodies* (1612; in Italian: *Discorso intorno alle cose*

che stanno in su l'acqua, o che in quella si muovono, "Discourse on Bodies that Stay Atop Water, or Move in It")

- *History and Demonstration Concerning Sunspots* (1613; in Italian: *Istoria e dimostrazioni intorno alle macchie solari*; work based on the *Three Letters on Sunspots, Tre lettere sulle macchie solari*, 1612)

- "Letter to the Grand Duchess Christina" (1615; published in 1636)

- "Discourse on the Tides" (1616; in Italian: *Discorso del flusso e reflusso del mare*)

- *Discourse on the Comets* (1619; in Italian: *Discorso delle Comete*)

- *The Assayer* (1623; in Italian: *Il Saggiatore*)

- *Dialogue Concerning the Two Chief World Systems* (1632; in Italian:

Dialogo sopra i due massimi sistemi del mondo)

- *Discourses and Mathematical Demonstrations Relating to Two New Sciences* (1638; in Italian: *Discorsi e Dimostrazioni Matematiche, intorno a due nuove scienze*)

ABOUT THE AUTHOR

ROBERT STAWELL BALL

(July 1, 1840 – November 25, 1913)

Sir Robert Stawell Ball was an Irish astronomer who founded the screw theory.

He was the son of naturalist Robert Ball and Amelia Gresley Hellicar. He was born in Dublin.

Ball worked for Lord Rosse from 1865 to 1867. In 1867 he became Professor of Applied Mathematics at the Royal College of Science in Dublin. There he lectured on mechanics and published an elementary account of the science.

In 1874 Ball was appointed Royal Astronomer of Ireland and Andrews Professor of Astronomy in the University of Dublin at Dunsink Observatory.

Ball contributed to the science of kinematics by delineating the screw displacement:

When Ball and the screw theorists speak of screws they no longer mean actual cylindrical objects with helical threads cut into them but the possible motion of any body whatsoever, including that of the screw independently of the nut.

Ball's treatise *The Theory of Screws* (1876) is now in the public domain. His work on screw dynamics earned him in 1879 the

Cunningham Medal of the Royal Irish Academy.

In 1882 Popular Science Monthly carried his article "*A Glimpse through the Corridors of Time*". The following year it carried his two-part article on "*The Boundaries of Astronomy*".

Ball expounded the tides in *Time and Tide*: a Romance of the Moon In 1892 he was appointed Lowndean Professor of Astronomy and Geometry at Cambridge University at the same time becoming director of the Cambridge Observatory. He was a fellow of King's College, Cambridge.

In 1900 Cambridge University Press published *A Treatise on the Theory of Screws*. That year he also published *The Story of the Heavens*. Much in the limelight, he stood as President of the Quaternion Society. He was also President of the Mathematical Association in 1900.

In 1908 he published *A Treatise on Spherical Astronomy*, which is a textbook on astronomy starting from spherical trigonometry and the celestial sphere, considering atmospheric refraction and aberration of light, and introducing basic use of a generalised instrument.

His work *The Story of the Heavens* is mentioned in the "Ithaka" chapter of Ulysses. His lectures, articles and books (e.g. *Starland* and *The Story of the Heavens*) were mostly popular and simple in style.

He died in Cambridge and was buried at the Parish of the Ascension Burial Ground in Cambridge, with his wife Lady Francis Elizabeth Ball. Their children were: Frances Amelia, Robert Steele, William Valentine (later Sir), Mary Agnetta, Charles Rowan Hamilton, and Randall Gresley (later Colonel).

LECTURES

Ball became celebrated for his popular lectures on science. He gave an estimated 2500 lectures between 1875 and 1910 in towns and cities across Britain and Ireland. In 1892, 1898 and 1900 he was invited to deliver the Royal Institution Christmas Lecture. *Astronomy; Astronomy* and *Great Chapters from the Book of Nature.*

FEEDBACK

Now that you have read the book ...

Was it interesting?

Did you enjoy what you wanted to read?
Was there any room for improvement?

Let us know at:

http://www.diamondbooks.ca/feedback

Your feedback is highly appreciated.
Thank you!

- ROBERT STAWELL BALL

Would you like to buy a copy of 'ISSAC NEWTON?'

PLEASE VISIT:

http://www.diamondbooks.ca

Made in United States
Cleveland, OH
06 November 2026

47647260R00093